Science: a third level course

S343 INORGANIC CHEMISTRY

INTRODUCING THE TRANSITION ELEMENTS

The Open University

Prepared by an Open University Course Team

S343 Course Team

Course team chairs — *Stuart Bennett, Elaine Moore, Lesley Smart*

Authors
- Block 1 *David Johnson*
- Block 2 *Elaine Moore*
- Block 3 *Kiki Warr, with contributions from David Johnson*
- Block 4 *Stuart Bennett*
- Block 5 *Michael Mortimer*
- Block 6 *Ivan Parkin, with contributions from Dr M. Kilner, Professor K. Wade* (University of Durham) *and F. R. Hartley* (Royal Military College of Science, Shrivenham)
- Block 7 *Paul Walton* (University of York), *with contributions from Lesley Smart*
- Block 8 *Lesley Smart, with contributions from David Johnson, Kiki Warr and Elaine Moore*
- Block 9 *David Johnson*

Consultants
- Dr P. Baker (University College of North Wales)
- Dr R. Murray (Trent Polytechnic)

Course managers
- *Peter Fearnley*
- *Wendy Selina*
- *Charlotte Sweeney*

Editors
- *Ian Nuttall*
- *David Tillotson*

BBC
- *Andrew Crilly*
- *David Jackson*
- *Jack Koumi*
- *Michael Peet*

Graphic artists
- *Steve Best*
- *Janis Gilbert*
- *Andrew Whitehead*

Graphic designers
- *Josephine Cotter*
- *Sarah Hofton*

Assistance was also received from the following people:
- *George Loveday* (Staff Tutor)
- *Joan Mason*
- *Jane Nelson* (Staff Tutor)

Course assessor — *Professor J. F. Nixon* (University of Sussex)

The Open University, Walton Hall, Milton Keynes, MK7 6AA

First published 1989. Reprinted 1992, 2000, 2003

Copyright © 1989 The Open University

All rights reserved. No part of this publication may be reproduced, stored in a retrieval system or transmitted in any form or by any means, without written permission from the publisher or a licence from the Copyright Licensing Agency Limited. Details of such licences (for reprographic reproduction) may be obtained from the Copyright Licensing Agency Ltd of 90 Tottenham Court Road, London W1P 0LP.

Edited and designed by the Open University. Typeset by Santype International Limited, Salisbury, Wiltshire.

Printed in the United Kingdom by Martins the Printers Ltd, Berwick upon Tweed

ISBN 07492 5000 3

This text forms part of an Open University Level 3 course. If you would like a copy of Studying with the Open University, please write to the Course Reservations and Sales Centre, PO Box 724, The Open University, Walton Hall, Milton Keynes, MK7 6ZS, United Kingdom. If you have not enrolled on the Course and would like to buy this or other Open University material, please write to Open University Worldwide, The Berrill Building, Walton Hall, Milton Keynes, MK7 6AA, United Kingdom: tel. +44 (0)1908 858585, fax +44 (0)1908 858787, e-mail ouwenq@open.ac.uk. Alternatively, much useful course information can be obtained from the Open University's website http://www.open.ac.uk.

1.4

S343Block1i1.4

STUDY GUIDE FOR BLOCK 1

1 INTRODUCTION TO THE COURSE

2 WHAT ARE THE TRANSITION ELEMENTS?
2.1 What electronic configurations do the transition elements have?
2.2 Electronic configuration and oxidation state
2.3 Summary of Section 2

3 TITANIUM
3.1 Sources of titanium
3.2 Titanium dioxide
   3.2.1 Manufacture of titanium dioxide
   3.2.2 Titanium dioxide as a white pigment
3.3 Titanium tetrachloride
3.4 Titanium metal
   3.4.1 Titanium extraction
   3.4.2 Uses of titanium metal
3.5 Other compounds of titanium(IV)
3.6 Lower oxidation states of titanium
   3.6.1 Aqueous ions
   3.6.2 Colour and transition-metal compounds
   3.6.3 Lower halides of titanium
3.7 Ziegler–Natta catalysis
3.8 Summary of titanium chemistry

4 MANGANESE
4.1 Sources of manganese
4.2 Manganese metal
   4.2.1 Preparation of metallic manganese
   4.2.2 Manganese and the steel industry
4.3 Manganese halides
4.4 The oxidation states of manganese in aqueous solution
   4.4.1 Acid solution
   4.4.2 Alkaline solution
   4.4.3 General comments
4.5 Manganese dioxide and lower oxides
   4.5.1 Manganese dioxide and batteries
4.6 Manganese nodules
4.7 Potassium permanganate
4.8 Summary of manganese chemistry

5 COBALT
5.1 Sources of cobalt
5.2 Cobalt metal
   5.2.1 Preparation of metallic cobalt
5.3 Simple aqueous chemistry of cobalt
5.4 Cobalt halides
5.5 Compounds of cobalt and ammonia
   5.5.1 Testing Werner's theory
   5.5.2 The anatomy of the complex
   5.5.3 Geometric isomerism of cobalt(III) complexes
   5.5.4 Optical isomerism of cobalt(III) complexes
   5.5.5 Werner's work in a wider context
5.6 Some cobalt(II) complexes
5.7 Higher oxidation states of cobalt
5.8 Summary of cobalt chemistry

6 WIDENING THE PERSPECTIVE

7 THE ELEMENTAL STATE   40

8 THE OXIDATION STATE +2   41

    8.1 Aqueous ions   41
    8.2 Dihalides   42
        8.2.1 Structures of dihalides   43
        8.2.2 Radii of dipositive ions   43
    8.3 Some reactions within the dipositive oxidation state   45

9 THE OXIDATION STATE +3   46

    9.1 Aqueous ions   46
    9.2 Thermodynamic stability of dipositive and tripositive aqueous ions   48
        9.2.1 The redox potential $E^{\ominus}(M^{3+}|M^{2+})$ 48
        9.2.2 Solvent decomposition   50
    9.3 Thermodynamic stability and electronic configuration   51
    9.4 Trihalides   52
    9.5 Summary of Sections 8 and 9   54

10 FINAL COMMENTS ON OXIDATION STATES +2 AND +3   55

11 LOWER OXIDATION STATES   56

    11.1 Copper   56
    11.2 Summary of Section 11   58

12 HIGHER OXIDATION STATES IN AQUEOUS SOLUTION   58

    12.1 Chromium   58
    12.2 Manganese   59
    12.3 Iron   60
    12.4 Structural chemistry of the higher oxidation states of chromium, manganese and iron   60

13 SUMMARY OF OXIDATION-STATE PATTERNS IN THE FIRST TRANSITION SERIES   61

    13.1 The overall profile of oxidation states   62
    13.2 Summary of Section 13   64

14 SUMMARY OF BLOCK 1   64

15 VANADIUM: A FINAL CASE STUDY   65

APPENDIX 1 STANDARD REDOX POTENTIALS   66

APPENDIX 2 POTENTIAL DIAGRAMS   68

OBJECTIVES FOR BLOCK 1   69

SAQ ANSWERS AND COMMENTS   71

ACKNOWLEDGEMENTS   79

STUDY GUIDE FOR BLOCK 1

This Block is equivalent to two Units or two weeks' work. It provides a selective introduction to the chemistry of the first transition series. It starts with an introduction to the concept of a transition element, and a discussion of the significance of the concept of oxidation state. This is followed, in Sections 3–5.8, by a sketch of the chemistry of three first-row transition elements: titanium, manganese and cobalt. The end of Section 5.8 marks a convenient break-point, and you should aim to reach it by the end of your first week's study. Afterwards, in Sections 6–15, the discussion broadens beyond these three elements into a survey of some general properties of the whole of the first transition series.

In addition to the main text, there are two short videocassette sequences associated with Block 1, which demonstrate the nature and chemical reactions of some of the substances discussed in the text. Fuller details of the videocassette contents are to be found in the S343 *Audiovision Booklet*. The main text will advise you of any sequence on the videocassette which is especially relevant to your current studies.

Some Sections of the Block deal with crystal structures and stereochemistry. If your learning is impeded by difficulties in visualising three-dimensional structures from illustrations in the text, you may find it helpful to use your model kit to build models of some of the molecules discussed. This may be especially the case in Sections 5.5.2–5.5.4.

1 INTRODUCTION TO THE COURSE

SLC 1 You are about to start S343 Inorganic Chemistry. What kind of Course is it, and what are its aims? First of all, the subject matter: if you could glance through all nine Blocks, you would see that the Course is overwhelmingly about the transition elements, the lanthanides and the actinides. The typical elements do appear, but with nothing like the same intensity. And if you think about what you already know, you will not find this hard to understand: the typical elements were discussed extensively in a Second Level Course. This Course complements that treatment, and so gives you a more balanced knowledge of inorganic chemistry.

However, the choice of content has not just been motivated by the need for a respectable academic coverage of the chemical elements. The Course is also about research and industrial applications that are affecting the structure and funding of modern inorganic chemistry. In this respect, the transition elements, the lanthanides and the actinides are particularly influential.

For example, during the last 30 years, the synthesis of new organometallic compounds of the transition elements has uncovered novel types of chemical bonding, and supplied new catalysts for the manufacture of organic chemicals. Developments of this kind are studied in Block 6, which discusses organometallic chemistry.

Again, transition elements are essential micronutrients in the soil, and they appear in many important enzymes. In Block 7 you will see how the study of such things has led to an understanding of nitrogen fixation, which may one day have valuable agricultural applications.

Solid compounds of the transition elements and lanthanides have important electronic applications, for example in magnetic tape and new superconducting materials. You will get a flavour of this in Block 8 on solid state chemistry.

Finally, nations that run nuclear energy programmes maintain huge research establishments for the study of actinide elements, and their relatives, the lanthanides. Block 9 shows you how this research has been exploited in the chemistry of nuclear fuel cycles, including the latest type of nuclear fuel reprocessing plants.

These advanced topics appear in the second half of the Course, but before you can understand them, you need a grounding in basic transition-metal chemistry, and this is provided in Blocks 1–4. To begin with, in Block 1, you will meet some simple chemistry of the first-row transition elements, followed, in Block 2, by a treatment of bonding theories for transition-metal compounds and complexes.

Block 3 then reveals how thermodynamics, allied with the theory of Block 2, provides a deeper insight into the descriptive chemistry of Block 1.

Block 4 shows how the variety of transition-metal chemistry is enlarged by the introduction of new kinds of ligand, and how stereochemistry is just as important as it is in organic chemistry. The syntheses and structural investigation of the kinds of compound studied in the latter part of Block 4 are illustrated with a videocassette programme on laboratory techniques.

However, one such technique, the spectroscopic one of nuclear magnetic resonance (n.m.r.), is now so important that it merits a fuller treatment. So the gap between the basic chemistry of earlier Blocks, and the current research and applications of Blocks 6–9, is bridged by Block 5 on n.m.r.

That concludes our general introduction; you can now get started on Block 1. Here you will begin by revising the meaning of the term *transition element*, the electronic configurations of these elements, and the concept of oxidation state. Sections 3–5.8 are then largely given over to selected chemistry of three first-row transition elements: titanium, manganese and cobalt. Not only will this study increase your ability to solve chemical problems by acquainting you with new facts, but it will also be used to illustrate important concepts, and to provide examples of the socioeconomic and industrial importance of transition elements. With this background you will then be ready, in Sections 6–15, to broaden your perspective into a general survey of the chemistry of the entire first transition series.

Figure 1 Dmitri Mendeléev (1834–1907). On graduation, his fiery temper led him to quarrel with educational bureaucrats, who took revenge by posting him to a college that had been shut down. Continually at odds with the Russian establishment, he subsequently became a Professor of Chemistry in St Petersburg and a consultant on cheesemaking. He discovered the Periodic Law in 1869.

2 WHAT ARE THE TRANSITION ELEMENTS?

SFC 1 In the Foundation Course you saw how Dmitri Mendeléev (Figure 1) devised a short Periodic Table like that in Figure 2. Mendeléev used the term 'transition elements' to describe the Group VIII triads such as iron, cobalt and nickel. This was because some of their properties resembled those of elements such as chro-

Figure 2 A later version of Mendeléev's short form of the Periodic Table.

	Group 0	Group I A B	Group II A B	Group III A B	Group IV A B	Group V A B	Group VI A B	Group VII A B	Group VIII
Period 2	2 He	3 Li	4 Be	5 B	6 C	7 N	8 O	9 F	
Period 3	10 Ne	11 Na	12 Mg	13 Al	14 Si	15 P	16 S	17 Cl	
Period 4	18 Ar	19 K 29 Cu	20 Ca 30 Zn	21 Sc 31 Ga	22 Ti 32 Ge	23 V 33 As	24 Cr 34 Se	25 Mn 35 Br	26 Fe 27 Co 28 Ni
Period 5	36 Kr	37 Rb 47 Ag	38 Sr 48 Cd	39 Y 49 In	40 Zr 50 Sn	41 Nb 51 Sb	42 Mo 52 Te	43 Tc 53 I	44 Ru 45 Rh 46 Pd
Period 6	54 Xe	55 Cs 79 Au	56 Ba 80 Hg	57–71 * 81 Tl	72 Hf 82 Pb	73 Ta 83 Bi	74 W 84 Po	75 Re 85 At	76 Os 77 Ir 78 Pt
Period 7	86 Rn	87 Fr	88 Ra	89 Ac	90 Th	91 Pa	92 U		

* The rare earth elements
57 La   58 Ce   59 Pr   60 Nd   61 Pm   62 Sm   63 Eu   64 Gd   65 Tb   66 Dy   67 Ho   68 Er   69 Tm   70 Yb   71 Lu

mium and manganese that preceded them in the same row, whereas other properties resembled those of elements like copper that followed them at the beginning of the next row. It was thus possible to view the triads as a gradual transition between the elements at the end of one row and those at the beginning of the next. By contrast, other rows ended in a halogen, and the next then began with a noble gas*. In each of these cases, therefore, there was a discontinuous change in properties, which, if it could be called a transition was certainly not a gradual one.

Now turn to the modern long form of the Periodic Table, which appears on the back cover. In this version, Mendeléev's term 'transition elements' has been taken up with an enlarged meaning. The nature of the typical elements is defined by the structure of Periods 1, 2 and 3; the elements in Period 4 between the typical elements on the left-hand side of the Table and the typical elements on the right-hand side are called **transition elements**. They are the elements from scandium to zinc inclusive. In this Course, we also define the transition elements in Periods 5 and 6 as those that fall beneath the elements from scandium to zinc, so in these three Periods, there are 30 in all†.

2.1 What electronic configurations do the transition elements have?

The electronic configuration of the free argon atom, [Ar], is $1s^2 2s^2 2p^6 3s^2 3p^6$. In the succeeding atoms from potassium to zinc, electrons enter the 3d or 4s levels, and for these elements, both 3d and 4s electrons are called *valence electrons*. In the potassium and calcium atoms the valence electrons enter the 4s level, so the electronic configuration of calcium can be written $[Ar]4s^2$. At scandium the 3d level begins to fill. The resulting electronic configurations of the free atoms of the first-row transition elements are shown in column 2 of Table 1.

Table 1 Electronic configurations of the free atoms and dipositive ions of the first transition series

Element	Free atom	Free M^{2+} ion
Sc	$[Ar]3d^1 4s^2$	$[Ar]3d^1$
Ti	$[Ar]3d^2 4s^2$	$[Ar]3d^2$
V	$[Ar]3d^3 4s^2$	$[Ar]3d^3$
Cr	$[Ar]3d^5 4s^1$	$[Ar]3d^4$
Mn	$[Ar]3d^5 4s^2$	$[Ar]3d^5$
Fe	$[Ar]3d^6 4s^2$	$[Ar]3d^6$
Co	$[Ar]3d^7 4s^2$	$[Ar]3d^7$
Ni	$[Ar]3d^8 4s^2$	$[Ar]3d^8$
Cu	$[Ar]3d^{10} 4s^1$	$[Ar]3d^9$
Zn	$[Ar]3d^{10} 4s^2$	$[Ar]3d^{10}$

Note that there is a gradual filling of the 3d shell across the series. This filling is not quite regular, however, because at chromium and copper the population of the 3d level is increased by the acquisition of one of the two 4s electrons. Apart from zinc and copper, the free atoms have incomplete 3d shells. As we shall see, in many ways the chemistry of the transition elements is more easily related to the electronic configurations of free ions than to that of free atoms. The relevant ions have a charge of +2 or more (+1 in the case of copper), and their electronic configurations can be obtained by removing *first* the outer s electron(s) of the free atom, and *second*, the outer d electrons, until the total number of electrons

---

* Note that the noble gases appear on the right in the long form of the Periodic Table (back cover).

† Sometimes broader or narrower definitions are used. In Periods 6 and 7, the elements bridging the two sections of typical elements include the lanthanides and actinides, so some chemists also include them among the transition elements. However, others not only omit the lanthanides and actinides as we have done, but they leave out one or both of the scandium and zinc groups as well. This is because these two groups lack characteristics like colour and variety of oxidation state, which are typical of other transition elements.

removed is equal to the charge on the ion. For example, the electronic configurations of the dipositive ions obtained by applying this generalisation are shown in column 3 of Table 1. Notice that, unlike in the free atoms, there is now a smooth progression in the 3d electron population.

2.2 Electronic configuration and oxidation state

SLC 2

Throughout this Course, much use is made of oxidation states. This concept was introduced in a Second Level Course.

□   What is the oxidation state of iron in the compounds Fe_2O_3 and $FeCl_3$, and in the ion $[FeF_6]^{3-}$?

■   +3 in all three cases: oxygen and the halogens are assigned oxidation states of -2 and -1, respectively; for each species, the sum of the oxidation states must equal the overall charge.

SLC 3

In the Second Level Course, we made the important point that the oxidation state of an atom in a compound is in no sense a measure of the charge carried by the atom. Thus, when we classify together compounds or ions in the same oxidation state such as Fe_2O_3, $FeCl_3$, $FeBr_3$, $[FeCl_4]^-$ and $[FeF_6]^{3-}$, there is no suggestion that the iron atom in these species carries a charge of $+3$, or even the same charge. The charge would be $+3$ if the bonding were completely ionic, but it will be reduced below this value by varying degrees of covalency. Nevertheless, the electronic structures of the five species do have something in common.

To see what it is, consider the case of $FeCl_3$ and assume first that the bonding is wholly ionic. The compound will then be composed of Fe^{3+} and Cl^- ions.

□   What is the electronic configuration of the Fe^{3+} ion?

■ $[Ar]3d^5$; remove the two 4s electrons and then one 3d electron from the free iron atom.

The three electrons removed are transferred to the 3p orbitals of three chlorine atoms to form three Cl^- ions. This transfer leads to electrostatic bonding between the oppositely charged ions, so we can describe the three electrons transferred per iron atom plus the three electrons that they pair up with on chlorine as bonding electrons: the five 3d electrons of the Fe^{3+} ion are not involved in bonding.

Now suppose we allow some degree of covalency. Each pair of bonding electrons no longer resides exclusively on chlorine: it is partially shared with what was an Fe^{3+} ion.

□   Does this have any effect on the five 3d electrons on iron?

■   One's first reaction might be that the distribution of the bonding electrons between iron and chlorine will have little or no effect on the electrons that are not involved in bonding.

In Block 2, you will see that this is not so. Even in a purely ionic compound, the $3d^5$ shell has properties different from those that it has in the free ion. Nevertheless, whether the bonding is ionic or covalent, it is still possible to detect experimentally five outer electrons in compounds of iron in oxidation state $+3$. These five electrons reside mainly on iron. Their properties can be related to those of the five 3d electrons in the free Fe^{3+} ion, and the establishment of this link will be one of our most important tasks in Block 2. Chemists therefore say that in compounds of the same oxidation state, a particular transition element usually has the same number of d electrons: classification of the compounds of a particular transition element by oxidation state usually classifies them by the number of d electrons as well. This classification leads to other important relationships between compounds of a transition element in the same oxidation state.

A quick way of obtaining the number of d electrons that a transition element has in a particular compound is suggested by the example of $FeCl_3$ above. Firstly, compute the oxidation state of the transition element, and *for the purpose of the exercise*, think of this as a charge. The number of d electrons is that borne by the free ion with this charge. This method is quite satisfactory for all compounds with oxidation states of $+2$ or more, our main concern in Blocks 1–3.

2.3 Summary of Section 2

1 The electronic configurations of transition-metal ions with a charge of $+2$ or more can be found by removing first the outer s electrons from the electronic configurations of the free atoms, and second the appropriate number of outer d electrons to achieve that charge.

2 Classification of the compounds of a transition element by oxidation state also classifies them by the number of d electrons. For oxidation states of $+2$ or more, this number can be found by treating the oxidation state of the element as if it were a charge, and following the procedure in 1 above.

SAQ 1 The electronic configurations of the free chromium and cobalt atoms are $[Ar]3d^54s^1$ and $[Ar]3d^74s^2$, respectively. What are the configurations of the free ions Cr^{3+} and Co^{4+}?

SAQ 2 State the oxidation state of manganese and the number of 3d electrons associated with manganese in the following compounds and ions: (a) $MnCl_2$; (b) $[MnO_4]^{2-}$; (c) K_3MnF_6; (d) Mn_2O_7; (e) $Mn_2(SO_4)_3$; (f) $Mn(NO_3)_2$.

3 TITANIUM

Titanium is the ninth most abundant element (0.6 per cent by mass) in the Earth's crust, and the second most abundant transition element after iron. It occurs in Group IVA of Mendeléev's Periodic Table (Figure 2), and is classified with zirconium and hafnium both there and in the long form (back cover). The electronic configuration of the free atom is $[Ar]3d^24s^2$.

The most widely used titanium substances are the metal, the dioxide, TiO_2, and the tetrachloride, $TiCl_4$. The titanium metal industry has developed only since 1945, and has suffered large fluctuations in the demand for its products. The dioxide is the paint industry's most important white pigment. The tetrachloride is an important intermediate in the production of both the metal and the dioxide; it also plays a key role in a process that won its inventors the 1963 Nobel Prize for Chemistry.

3.1 Sources of titanium

There are two important commercial sources of the element: ilmenite, $FeTiO_3$, and rutile, TiO_2. Ilmenite is much the more plentiful, but because rutile is richer in titanium and lacks unwanted iron, it is the more desirable. Large rutile deposits occur in Australia and Sierre Leone. Countries rich in ilmenite are the United States, Norway, Canada and the Soviet Union. There are two commercial processes for the manufacture of chemically pure TiO_2, one for each of the two starting materials.

3.2 Titanium dioxide

Titanium dioxide occurs in three different crystalline forms, rutile, anatase and brookite, but here we shall only consider rutile, which is by far the most important. The tetragonal unit cell of the rutile structure is shown in Figure 3.

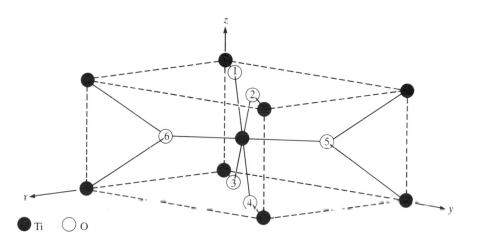

Figure 3 The unit cell of the rutile structure. It is tetragonal: if you look along the x and y axes, you see rectangles but if you look down the z axis, you see a square.

☐ What kind of coordination do titanium and oxygen have in this structure?

■ Each titanium is surrounded by six oxygens at the corners of an octahedron: oxygens 1, 2, 3 and 4 lie at the corners of a square around the central titanium; oxygens 5 and 6 lie on a line perpendicular to this square. Examination of oxygen 5 (or 6) shows that each oxygen is surrounded by three titaniums at the corners of a triangle. Therefore the coordination number of titanium with respect to oxygen is 6, and that of oxygen with respect to titanium is 3.

☐ How are these coordination numbers reflected in the formula TiO_2?

■ As there are two oxygens for each titanium, the coordination number of titanium must be twice that of oxygen.

Titanium dioxide is a white solid. When strongly heated in air, the compound first loses a little oxygen and then melts at about 1 850 °C. It can be dissolved with some difficulty in hot, fairly concentrated HCl, H_2SO_4 and $HClO_4$ to give colourless solutions containing an aqueous titanium(IV) species. The exact nature of this species is uncertain, but it is definitely not the aqueous monatomic ion, Ti^{4+}(aq), which is unknown. The formulations $[Ti(OH)_2]^{2+}$(aq) and TiO^{2+}(aq) have both been proposed and here we shall assume the latter. This implies that the unknown Ti^{4+}(aq) is unstable with respect to the hydrolysis reaction,

$$Ti^{4+}(aq) + H_2O(l) = TiO^{2+}(aq) + 2H^+(aq) \qquad 1$$

TiO_2 is also quite resistant to attack by concentrated alkaline solutions, but it dissolves in molten alkalis like KOH and $Ba(OH)_2$ to give mixed oxides such as K_2TiO_3 and $BaTiO_3$.

3.2.1 Manufacture of titanium dioxide

1 The sulphate process
The **sulphate process** uses ilmenite as the starting material. One modern version (Figure 4) is operated continuously by agitating ilmenite with H_2SO_4 (*ca.* 6 mol l^{-1}) at temperatures of 60–140 °C. This gives a solution containing iron(II) and titanium(IV):

$$FeTiO_3(s) + 4H^+(aq) = Fe^{2+}(aq) + TiO^{2+}(aq) + 2H_2O(l) \qquad 2$$

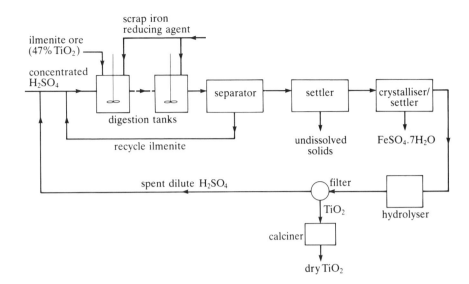

Figure 4 Flow chart of the sulphate process for manufacturing TiO$_2$ from ilmenite.

During this stage, some scrap iron is also added to reduce any iron(III) in the ilmenite to iron(II):

$$Fe(s) + 2Fe^{3+}(aq) = 3Fe^{2+}(aq) \qquad 3$$

This step is essential to avoid contamination of the TiO$_2$ product with iron(III), which would mar its whiteness.

Insoluble material is then removed, and the liquid cooled, which causes iron(II) to crystallise as the hydrated sulphate, FeSO$_4$.7H$_2$O. The liquid is separated from the sulphate in the crystalliser/settler and diluted with water to hydrolyse the dissolved titanium(IV) and precipitate hydrated TiO$_2$:

$$TiO^{2+}(aq) + (n+1)H_2O(l) = TiO_2 \cdot nH_2O(s) + 2H^+(aq) \qquad 4$$

Formulae of the type (oxide.nH$_2$O) are used when we are ignorant of how much water is incorporated into the structure. They imply that when the solid is filtered off and 'calcined' (that is, heated), the water is driven off, in this case, leaving pigment-grade TiO$_2$:

$$TiO_2 \cdot nH_2O(s) = TiO_2(s) + nH_2O(g) \qquad 5$$

The sulphuric acid filtrate, now about 2 mol l^{-1}, is recycled to the beginning, where, following the addition of concentrated acid, it can be used to dissolve more ilmenite.

2 The chloride process

The **chloride process** uses rutile mineral as the starting material, and then purifies it. This method was introduced by Du Pont in the United States in 1958. The rutile is first converted to the tetrachloride by heating it with coke and chlorine at about 950 °C

$$TiO_2(s) + 2C(s) + 2Cl_2(g) = TiCl_4(g) + 2CO(g) \qquad 6$$

The tetrachloride is condensed to a liquid, and distilled to remove impurities. When the vapour reacts with oxygen or air in a flame at about 1 500 °C, chlorine gas and a cloud of fine particles of TiO$_2$ are produced:

$$TiCl_4(g) + O_2(g) = TiO_2(s) + 2Cl_2(g) \qquad 7$$

The chlorine can be recycled to make more tetrachloride.

During the 1980s, approximately 3 million tonnes of TiO$_2$ were produced annually, about two-thirds by the sulphate route and one-third by the chloride method. In the United States, however, the chloride process accounts for about 75 per cent of TiO$_2$ output.

The chloride process is technically difficult: because it uses high temperatures, the corrosion problems are considerable, and the use of the easily hydrolysed TiCl$_4$ requires strict exclusion of water. Another disadvantage is that the starting material, rutile, is less common and more expensive than ilmenite. On the other hand this process is compact, the chlorine by-product can be recycled, and the distillation of the TiCl$_4$ leads to a particularly pure product. However, the technology of the sulphate process is much simpler, and ilmenite is cheap: the major disadvantage is that much sulphuric acid is consumed in producing an unwanted by-product, FeSO$_4$.7H$_2$O.

SAQ 3 A simplified flow chart for the industrial synthesis of TiO$_2$ by the chloride process is shown in Figure 5. Identify substances A–E and write down equations for reactions 1 and 2.

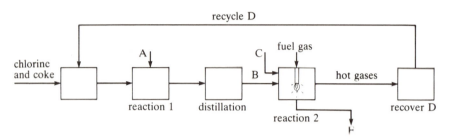

Figure 5 Flow chart of TiO$_2$ production by the chloride process (to be used in SAQ 3).

SAQ 4 Ilmenite is cheaper than rutile, and hot coke and chlorine will convert it to TiCl$_4$:

$$\text{FeTiO}_3(s) + 3\text{C}(s) + 3\text{Cl}_2(g) = \text{TiCl}_4(g) + \text{FeCl}_2(l) + 3\text{CO}(g) \qquad 8$$

Why is ilmenite not used as the starting material in the chloride process for manufacturing TiO$_2$?

3.2.2 Titanium dioxide as a white pigment

The paint industry is the biggest consumer of TiO$_2$. The pigment in a typical exterior white paint consists of pure rutile eked out with other white materials such as clay or mica. Fine particles of this pigment are suspended in a liquid vehicle such as linseed oil. During 'drying', the oil is attacked by atmospheric oxygen and forms an elastic film. Within this film, the TiO$_2$ particles are suspended.

Unlike the old white lead pigments, TiO$_2$ is non-toxic, but its main value as a pigment derives from its **hiding power**—the capacity to conceal a surface on which it is painted. This hiding power is a consequence of the very high refractive index of titanium dioxide. From the Foundation Course you know how light can be deflected at the boundary of two media of different refractive index, either by refraction or internal reflection. Because the refractive index of rutile (2.76) is much larger than that of the dried linseed oil (1.48) in which it is suspended, such deflections are unusually large.

SFC 2

Figure 6 illustrates how refraction by three spherical TiO$_2$ particles operating in sequence can throw a light ray back before it reaches the underlying surface.

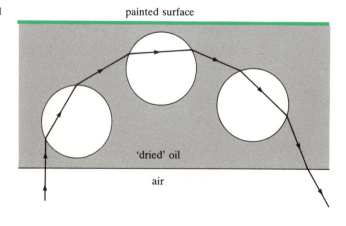

Figure 6 Three TiO$_2$ particles, suspended in an elastic film of 'dried' linseed oil, act in succession to reverse the path of the light ray. The effect is large because the refractive index of TiO$_2$ is much larger than that of the surrounding film.

Figure 7 compares the hiding power of paints containing 20 per cent by volume of different white pigments. Notice that when, as in the cases of white clay and calcium carbonate, the refractive index of the pigment is close to that of the surrounding film, the surface beneath is clearly visible. With anatase, it is almost completely obscured, and with rutile entirely so.

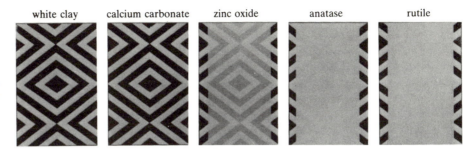

Figure 7 The hiding power (opacity) of paints containing 20 per cent by volume of white pigments of increasing average refractive index: white clay (1.56), calcium carbonate (1.57), zinc oxide (1.99), anatase (2.52) and rutile (2.76). The refractive index of the surrounding 'dried' film is 1.48. The traditional, but toxic, pigment, white lead, has a refractive index of 2.0.

3.3 Titanium tetrachloride

The manufacture of $TiCl_4$ by heating rutile with coke and chlorine at about 950 °C was covered in Section 3.2.1 (equation 6). $TiCl_4$ condenses to a colourless liquid with a boiling temperature of 136 °C. In moist air it smokes violently because of its ready hydrolysis, and with water the ultimate product is hydrated TiO_2:

$$TiCl_4(l) + 2H_2O(l) = TiO_2(s) + 4HCl(aq) \qquad 9$$

Titanium tetrachloride is an important polymerisation catalyst (see Section 3.7). It is also a crucial intermediate in the manufacture of pure TiO_2 (Section 3.2.1) and in the extraction of titanium metal, to which we now turn.

3.4 Titanium metal

At room temperature, titanium metal has the hexagonal close-packed structure encountered in a Second Level Course.

SLC 4

Table 2 Titanium compared with tin and aluminium

Metal	Outer electronic configuration	Density $\dfrac{}{g\,cm^{-3}}$	Melting temperature $\dfrac{}{°C}$	$\dfrac{\Delta H_f^\ominus(M, g)}{kJ\,mol^{-1}}$	Tensile strength $\dfrac{}{kg\,cm^{-2}}$
titanium	$3d^2 4s^2$	4.5	1 660	472	4 000
tin	$5s^2 5p^2$	7.3	232	302	220
aluminium	$3s^2 3p^1$	2.7	660	324	1 050

Transition metals are especially useful because they have greater mechanical strength and higher melting temperatures than their typical metal counterparts. Table 2 compares some properties of titanium with those of tin and aluminium. The Table includes values of $\Delta H_f^\ominus(M, g)$, the standard enthalpy change for the atomisation of the metal:

$$M(s) = M(g) \qquad 10$$

It also contains values for the tensile strength, the maximum load that a given cross-section of the metal can bear. This depends significantly on the way the specimens are prepared; the values in Table 2 are for samples of like origin.

Notice that the melting temperature and enthalpy of atomisation are significantly higher for the transition metal, which shows that the atoms are more tightly bound together. The comparison of titanium with tin is especially striking. Both the atoms have four outer electrons, but in titanium there are two d electrons, whereas in tin there are two p electrons. In Section 7, you will see further evidence that it is the presence of d electrons that leads to the strong binding forces in transition metals. This strong binding also plays a part in the higher tensile strengths.

From a commercial standpoint, the interesting comparison in Table 2 is that between titanium and aluminium. Both elements have excellent corrosion resistance, and for a similar reason, namely their ability to form tight oxide films on the metal surface. However, titanium has a melting temperature 1 000 °C higher than that of aluminium, and a tensile strength that is 3 or 4 times as great. Although the higher density of titanium is a structural disadvantage, the difference is not enough to offset the transition metal's greater strength. Moreover, titanium ores are plentiful. Yet in 1984, world production of aluminium metal was about 16 million tonnes, whereas that of titanium was well under 200 000 tonnes. The reason is to be found in the method of extraction.

3.4.1 Titanium extraction

Titanium cannot be obtained by heating the oxide with carbon because it forms carbides such as TiC. Instead, $TiCl_4$ is reduced with either sodium or magnesium. In the **Kroll process**, liquid $TiCl_4$ is dripped on to molten magnesium at 700–900 °C in a stainless steel vessel filled with argon gas:

$$2Mg(l) + TiCl_4(l) = Ti(s) + 2MgCl_2(l) \qquad 11$$

The molten magnesium chloride is tapped off. The titanium formed has a sponge-like appearance. Residual magnesium metal and magnesium dichloride are removed either by distilling them off in a vacuum, or by washing the sponge with dilute acids. The sponge can then be consolidated into ingots by melting it under vacuum in an electric arc. Note that this process uses an expensive reducing agent (magnesium) and is not continuous—it uses successive batches of $TiCl_4$, a substance that has the disadvantage of being hydrolytically unstable.

Do the following SAQ now.

SAQ 5 (*revision*) How does the extraction of aluminium differ in these respects?

The situation could be transformed by a continuous electrolytic process starting from TiO_2, but so far a workable method has not been discovered. A partial solution, which has reached the pilot plant stage, is an electrolytic method in which $TiCl_4$ is added to a molten mixture of LiCl and KCl under argon at 520 °C (Figure 8). The reduction occurs in two stages at steel cathodes. The $TiCl_4$ is first reduced to Ti^{2+} ions, which pass into the melt:

$$TiCl_4(l) + 2e^- = Ti^{2+}(melt) + 4Cl^-(melt) \qquad 12$$

Then there is further reduction to titanium sponge:

$$Ti^{2+}(melt) + 2e^- = Ti(s) \qquad 13$$

At the graphite anode, chlorine is produced:

$$4Cl^-(melt) = 2Cl_2(g) + 4e^- \qquad 14$$

The anode is screened by a diaphragm to prevent the chlorine from oxidising the reduced titanium. The cathode is periodically lifted, stripped of its titanium sponge deposit, and lowered back into the cell without disturbing the argon atmosphere.

It remains to be seen whether this method will replace the use of sodium or magnesium as a reducing agent.

3.4.2 Uses of titanium metal

Because extraction is expensive, the consumption of titanium does not approach that of aluminium. Owing to its good corrosion resistance, some titanium is used in chemical plant, but the principal outlets are in the aircraft industry, where the metal's properties make it peculiarly attractive. Even here, however, cost severely restricts consumption.

In Table 3 we compare three alloys that are widely used in the aircraft industry. The numbers before the symbols for the minor elements are the percentages by mass of these metals. Thus, Ti–6Al–4V contains 90 per cent Ti, 6 per cent Al and

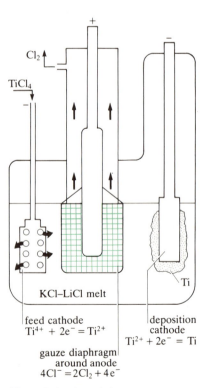

Figure 8 An electrolytic method for the extraction of titanium from $TiCl_4$.

4 per cent V. Notice first how much alloying raises the tensile strength above the values for unalloyed titanium and aluminium in Table 2. Then note that the strength: density ratio of the titanium alloy greatly exceeds those of stainless steel and its aluminium-based rival. This is the first reason why titanium alloys are valued by the aircraft industry. Thus, the discs, blading and carriers on the forward end of modern jet engines like the Rolls-Royce RB–211 are made of titanium alloys.

Table 3 Properties of three alloys that are put to structural use in the aircraft industry

Alloy	Tensile strength $\dfrac{}{\text{kg cm}^{-2}}$	Density $\dfrac{}{\text{g cm}^{-3}}$	Strength: density ratio $\dfrac{}{\text{cm}}$
Al–4Cu–2Mg–1Mn	4 900	2.77	1.77×10^6
Ti–6Al–4V	12 270	4.43	2.77×10^6
stainless steel	12 700	7.70	1.65×10^6

How about the airframe itself? Here the attractions of titanium increase with the speed of the aircraft. At very high speeds, the friction of the airstream raises the surface temperature of the airframe to values where conventional aluminium alloys are unusable. Figure 9 shows an aircraft that is made largely of Ti–6Al–4V. It has a top speed of about 2 000 mph at 80 000 feet. Under these conditions, the temperature of the wing's leading edge is about 425 °C, which is hotter than the average soldering iron; the rest of the surface is at 230–260 °C.

The virtues of titanium as a structural metal at these temperatures are apparent from Figure 10. This shows how the strength–density ratios of Table 3 change with temperature. Notice the steep fall in the efficiency of conventional aluminium alloys above 150 °C, the maximum temperature generated by aerodynamic heating at about Mach 2. At speeds near Mach 3, titanium rather than aluminium alloys must be used.

Figure 9 The Lockheed SR–71A Blackbird is the official holder of the world airspeed record (1987), 2 193 mph; the airframe is composed mainly of titanium.

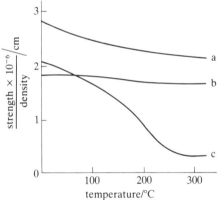

Figure 10 Strength: density ratios of alloys used in airframes plotted against temperature: a, titanium alloy; b, stainless steel; c, aluminium alloy (all solution treated and aged by 10 000 hours of exposure at particular temperatures).

Unfortunately for the titanium industry, this is not a serious problem for civil aircraft designers. The fastest operational airliner is the Anglo-French Concorde, and its design speed was set at Mach 2.2 to permit the use of an aluminium airframe. Its projected competitor, the Boeing SST was a Mach 3 aircraft containing over 60 tonnes of titanium, but it was cancelled on grounds of cost in 1971, causing a major recession in the titanium industry.

Thus, at present, it is mainly in military aircraft that titanium is to be found. More diverse uses are needed if the titanium industry is to become more secure and more stable.

SAQ 6 In the manufacture of aluminium, the metal can be obtained in a consolidated form, such as ingots, directly from the extraction cell. In titanium manufacture, a further step is necessary. What property is responsible for this difference between the two metals? Why does it also make the operation of an economical, continuous extraction process more difficult for titanium than for aluminium?

3.5 Other compounds of titanium(IV)

The +4 oxidation state is the highest oxidation state of titanium, a result that we might have anticipated from the position of the element in Group IVA of Mendeléev's Table.

□ How can this be related to the electronic configuration of the titanium atom?

■ The highest oxidation state is equal to the number of outer electrons.

The +4 oxidation state of titanium is also the commonest, an observation that is consistent with its presence in TiO_2 and $TiCl_4$, the two most important titanium compounds. $TiCl_4$, of course, is just one of the titanium tetrahalides, and some properties of all four are shown in Table 4. The preparation of $TiCl_4$ was described in Section 3.2.1; the other halides are conveniently prepared by heating titanium metal in the presence of the appropriate halogen, although for TiF_4, an especially favoured alternative is the reaction of the tetrachloride with anhydrous HF:

$$TiCl_4(l) + 4HF(l) = TiF_4(s) + 4HCl(g) \qquad 15$$

Table 4 Properties of titanium(IV) halides

Substance	Appearance at 25 °C	Melting temperature/°C	Boiling temperature/°C
TiF_4	white solid	*	284
$TiCl_4$	colourless liquid	−24	136.4
$TiBr_4$	orange solid	39	233
TiI_4	dark-brown solid	150	377

* TiF_4 sublimes without melting at the 'boiling temperature' quoted.

The liquid chloride, and the solid bromide and iodide contain discrete tetrahedral TiX_4 molecules.

SAQ 7 What evidence is there in Table 4 to suggest that this is not the case in solid TiF_4?

As noted in Section 3.2, TiO_2 dissolves in hot concentrated HCl, H_2SO_4 and $HClO_4$; on dilution it yields solutions containing an aqueous titanium(IV) species whose exact nature is uncertain, but which we describe here as TiO^{2+}(aq). From the titanium(IV) sulphate solution, the solid compound $TiOSO_4 \cdot H_2O$ can be recovered. However, X-ray studies show that this does *not* contain TiO^{2+} units; rather, the compound contains infinite chains of titanium and oxygen atoms (Figure 11). The overall coordination around each titanium is octahedral, the remaining four sites being occupied by four oxygen atoms, one being part of a water molecule, and the other three being part of three different sulphate groups.

SAQ 8 Figure 11 shows just one chain in $TiOSO_4 \cdot H_2O$. The different chains are held together by bridging sulphate groups. How many chains does any one sulphate group link together?

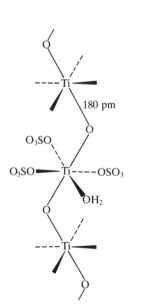

Figure 11 $TiOSO_4 \cdot H_2O$ contains infinite Ti—O chains. Each titanium atom is surrounded by six oxygens: two are in the chain, one is in a water molecule and three are in different sulphate groups.

3.6 Lower oxidation states of titanium

Compounds or aqueous ions of the +4 oxidation state of titanium can sometimes be reduced to oxidation states +2 and +3.

3.6.1 Aqueous ions

If a solution of TiO_2 in concentrated sulphuric acid is diluted with water and then heated with zinc, it turns purple. The purple colour is that of the $Ti^{3+}(aq)$ ion; titanium(IV) has been reduced to titanium(III):

$$Zn(s) + 2TiO^{2+}(aq) + 4H^+(aq) = Zn^{2+}(aq) + 2Ti^{3+}(aq) + 2H_2O(l) \qquad 16$$

Atmospheric oxygen slowly reoxidises $Ti^{3+}(aq)$ back to titanium(IV), so the solution is best preserved by keeping it under an unreactive gas such as nitrogen or argon. Attempts to reduce $Ti^{3+}(aq)$ further to $Ti^{2+}(aq)$ are unsuccessful even when very powerful reducing agents are used.

3.6.2 Colour and transition-metal compounds

Compounds of the transition elements are often coloured, and this is one of their most striking characteristics. Compounds are coloured when they absorb light in the visible region of the spectrum, a range of wavelength that runs approximately from 7×10^{-5} cm down to 4×10^{-5} cm. You saw at Second Level that radiation is often specified by its wavenumber, the reciprocal of its wavelength, because this is a measure of energy. The visible region then runs from about 14 000 to 25 000 cm^{-1} (see Plate 1.1 of S343 *Colour Sheet 1*). A solution of $Ti^{3+}(aq)$ is coloured because of a single broad absorption band with a peak at about 20 000 cm^{-1} in the visible spectrum (Figure 12). This absorbs particularly strongly near the green–blue boundary, transmitting much red light laced with the high-energy end of the blue; it therefore appears purple.

SLC 5

CS

SFC 3

As you know from the Foundation Course, light is absorbed by substances when its energy is used to bring about electronic transitions. Colour in transition-metal compounds is associated with two types of electronic transition, one of which is illustrated by $Ti^{3+}(aq)$.

□   What electronic configuration do you associate with titanium(III) substances?

■ $[Ar]3d^1$, the configuration of the Ti^{3+} ion.

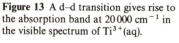

Figure 12 The visible absorption spectrum of $Ti^{3+}(aq)$ plotted against the wavenumber (reciprocal wavelength) of the radiation.

There are five 3d orbitals, and in the gaseous Ti^{3+} ion they all have the same energy. But this is not so in compounds or aqueous ions of titanium(III); in Block 2 we shall introduce a theory in which the 3d orbitals of the transition element are modified by compound formation and take on different energies from each other. It is then possible to observe electronic transitions in which an electron shifts from one d orbital to a d orbital of higher energy. Such a shift is known as a **d–d transition**. In transition-metal compounds, it so happens that the energies of d–d transitions often occur in the visible region of the spectrum. Look at Figure 13. In $Ti^{3+}(aq)$, there are two states with the configuration $[Ar]3d^1$, a ground state in which the 3d electron occupies a low-energy 3d orbital, and an excited state in which it occupies a high-energy 3d orbital. The excited state is about 20 000 cm^{-1} above the ground state, so visible light can incite the transition between the two, and the spectrum of the solution contains an absorption band with a peak at this energy.

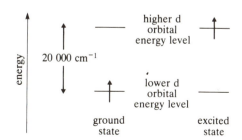

Figure 13 A d–d transition gives rise to the absorption band at 20 000 cm^{-1} in the visible spectrum of $Ti^{3+}(aq)$.

□   Now turn to Table 4. Why does it suggest that d–d transitions are not the only source of colour in transition-metal compounds?

■ Titanium(IV) bromide and iodide are coloured, yet there are no 3d electrons associated with the oxidation state titanium(IV).

The colours of TiBr$_4$ and TiI$_4$ are caused by **charge-transfer** transitions. In these transitions, an electron in an orbital predominantly located on the halogen is excited into an orbital located primarily on titanium: the *charge* of the electron is *transferred* from one atom to another rather than staying on the same site as in d–d transitions. Again, you will learn more about charge-transfer transitions in Block 2.

It is the *combination* of d–d and charge-transfer transitions, often occurring in the visible region, that makes transition-element compounds so colourful. Charge-transfer transitions are not excluded for compounds of the typical elements: they do occur, and sometimes generate colour, as in the yellow SnI$_4$, which you made in the Foundation Course. But in transition-metal compounds, charge-transfer and d–d transitions both occur, yielding an especially colourful chemistry.

SFC 4

3.6.3 Lower halides of titanium

All four titanium tetrahalides can be reduced to trihalides by heating them with the appropriate amounts of titanium metal at temperatures in the range 500–800 °C. The trihalides are all coloured solids. For example, TiF$_3$ is blue and TiCl$_3$ is violet. The visible absorption spectrum of both solid halides contains a single broad band like that in the spectrum of Ti^{3+}(aq) (Figure 12).

SAQ 9 In the spectrum of TiF$_3$, does the absorption band occur at lower or higher energy than in Ti^{3+}(aq)?

When TiCl$_3$, TiBr$_3$ and TiI$_3$ are heated under vacuum at 400–500 °C, they disproportionate, the tetrahalide vapour distilling off to leave solid black dihalides:

$$2\text{TiX}_3(s) = \text{TiX}_2(s) + \text{TiX}_4(g) \qquad 17$$

This reaction does not occur when TiF$_3$ is used, and all attempts to prepare TiF$_2$ have proved unsuccessful. If TiCl$_2$, TiBr$_2$ or TiI$_2$ is added to water or dilute acid, violent evolution of hydrogen occurs, and a reddish-purple solution is formed.

Do the following SAQ now.

SAQ 10 Write an equation for the reaction of TiCl$_2$ with dilute acid. How does it corroborate the observation of Section 3.6.1 that all attempts to produce Ti^{2+}(aq) have been unsuccessful?

SAQ 10 and its answer, together with the difficulties encountered in trying to prepare TiF$_2$, illustrate a general characteristic of titanium(II) compounds: they tend to be easily oxidised to titanium(III) or titanium(IV) compounds.

SLC 6

The structures of the titanium dihalides and trihalides are conveniently discussed together. TiCl$_2$, TiBr$_2$ and TiI$_2$ have the CdI$_2$ layer structure, which was discussed in a Second Level Course. It consists of a succession of three-decked layers, in which titanium atoms are sandwiched between two decks of halogen atoms. Each titanium is octahedrally coordinated by six halogens, three in the deck above, and three in the deck below. Figure 14 shows a view of one of these three-decked layers from above.

TiCl$_3$ and TiBr$_3$ both crystallise with a layer structure that bears a close relationship to that of the dihalides. The plan view of an individual layer equivalent to Figure 14 is shown in Figure 15. Again there is a three-decked layer, with titanium atoms sandwiched between two decks of halogen atoms, and again, each titanium is octahedrally coordinated by halogen.

□ What can you say about the disposition of the halogens in Figures 14 and 15?

- Ti in plane of paper
- halide above plane of paper
- halide below plane of paper

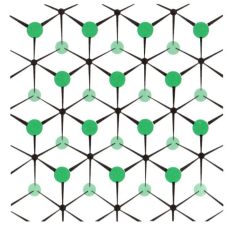

Figure 14 A view (from above) of one of the three-deck layers that make up the structure of the titanium dihalides. A layer of titanium atoms in the plane of the paper is sandwiched between halide layers above and below the plane, each titanium being octahedrally coordinated by halide. Such layer structures are similar to those of $CdCl_2$ and CdI_2, which were studied at Second Level.

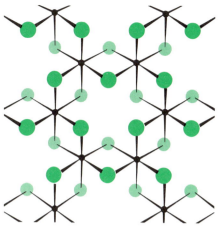

Figure 15 A view (from above) of one of the three-deck layers that make up the structure of $TiCl_3$ and $TiBr_3$. As in Figure 14, a titanium layer is sandwiched between halide layers, and titanium is in octahedral coordination. The halide layers are very similar to those in Figure 14, but within the titanium layer, the atoms are much less densely arranged.

■ They are identical; the difference between the two kinds of layer lies in the density of titanium atoms *within* the central deck.

SAQ 11 Figure 16 shows the arrangement of just the titanium atoms in the central deck of the $TiCl_2$ layer in Figure 14. They lie at the corners and centres of juxtaposed hexagons. The removal of the titaniums at the centres of the hexagons then results in the arrangement of titanium atoms in the central deck of the three-decked layer in the $TiCl_3$ structure of Figure 15. Prove that such removal eliminates one-third of the titanium atoms in Figure 16, and show that this proportion is consistent with the formulae $TiCl_2$ and $TiCl_3$.

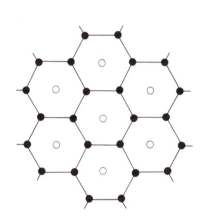

Figure 16 The arrangement of titanium atoms in the central deck of the three-deck layer of $TiCl_2$. When the atoms marked by open circles are removed, the arrangement in the central deck of the three-deck layer structure of $TiCl_3$ results.

The halides of titanium illustrate the point that the transition metals more frequently display strings of oxidation states differing by one than do the typical metals. Thus, titanium forms the chlorides $TiCl_2$, $TiCl_3$ and $TiCl_4$. By contrast, typical elements commonly display oxidation states that differ by multiples of two. For example, lead, like titanium, forms a liquid tetrachloride, but when this is reduced, $PbCl_3$ is not an observed product. Instead, reduction yields the white solid, $PbCl_2$, a reflection of the fact that lead chemistry is dominated by the oxidation states $+2$ and $+4$. Because transition-metal oxidation states often differ by one, a transition metal packs more oxidation states into its chemistry than does its typical-metal counterpart. In this particular respect, therefore, the chemistry of the transition metals shows more variety than that of the typical metals.

3.7 Ziegler–Natta catalysis

The first industrial process for the manufacture of polyethene was introduced by ICI in 1939. Ethene was heated to 200 °C at pressures of 1 000–2 000 atmospheres:

$$n H_2C=CH_2 \longrightarrow (CH_2-CH_2-)_n \qquad 18$$

Clearly, a process operating at lower temperatures and pressures was desirable, and in the 1950s it was provided through the efforts of a Swiss chemist, Karl Ziegler, and an Italian, Giulio Natta. The secret is a catalyst that can be made by dissolving the compound triethylaluminium, of empirical formula $Al(CH_2CH_3)_3$, in heptane, and adding $TiCl_4$. This results in reduction of $TiCl_4$ to a brown suspension of $TiCl_3$ and the polymerisation of ethene then takes place on the surface of the brown solid.

A typical industrial process operates at 50–100 °C, and 10 atmospheres. Other similar polymers such as polypropene, $(CH(CH_3)-CH_2-)_n$, can be made in the same system, as much as a quarter of a tonne of polymer being produced per gram of titanium in the catalyst! In 1963, Ziegler and Natta's discoveries were acknowledged by the Nobel Prize for Chemistry.

The mechanism of Ziegler–Natta catalysis is very interesting, but we defer discussion of it until Block 6 on organometallic chemistry has introduced you to the concepts that you will need to understand it.

3.8 Summary of titanium chemistry

1 The highest oxidation state of +4 is equal to both the Group number in Mendeléev's Table, and to the number of outer electrons in the titanium atom.

2 If we confine ourselves to halides, oxides, sulphides and aqueous ions, the observed oxidation states are +2, +3 and +4, the +2 state being very easily oxidised.

3 The colour of titanium compounds is due to both charge-transfer and d–d transitions, the latter only being apparent in oxidation states less than four, when d electrons are present.

4 Titanium metal is corrosion resistant, and has a high strength:density ratio, but the expensive and technically difficult batch process that is used to make it precludes wide use of the metal.

5 The use of TiO_2 as the paint industry's principal white pigment can be explained mainly by its high refractive index.

6 The known dihalides of titanium crystallise with the CdI_2 layer structure. Removal of one-third of the titanium atoms from these structures yields layer structures for $TiCl_3$ and $TiBr_3$. Throughout, titanium remains octahedrally coordinated by halogens.

7 Ethene undergoes polymerisation to polyethene at the surface of solid $TiCl_3$ suspended in a solution of $Al(CH_2CH_3)_3$ in heptane.

Some of the more important reactions of titanium and its compounds are summarised in Figure 17 and are shown in the first sequence on Videocassette 1.

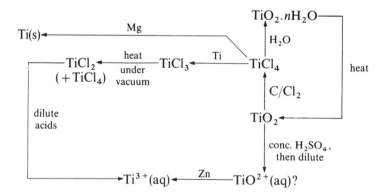

Figure 17 Some important reactions of titanium discussed in the text.

SAQ 12 Tin occurs in Group IVB of Mendeléev's Periodic Table (Figure 2) and titanium in Group IVA. Which of the following properties of tin and its compounds encourage the classification of tin with titanium in such a Table, and which do not?

(a) The highest oxide and chloride of tin are SnO_2 and $SnCl_4$, respectively.

(b) SnO_2 has the rutile structure.

(c) At room temperature, $SnCl_4$ is a colourless liquid, which fumes in moist air due to hydrolysis.

(d) Metallic tin melts easily in a household fire.

(e) The halides, oxides and aqueous ions of tin contain tin in just two oxidation states: +2 and +4.

4 MANGANESE

Manganese is the twelfth most abundant element (0.08 per cent by mass) in the Earth's crust, and the third most abundant transition element after iron and titanium. It occurs in Group VIIA of Mendeléev's Periodic Table, and is classified with technetium and rhenium. The electronic configuration of the free atom is $[Ar]3d^5 4s^2$.

As you will see, the industrially important manganese substances are alloyed manganese metal and the dioxide, MnO_2. Academically, manganese is interesting because of the variety and range of its oxidation states: all oxidation states from +2 to +7 inclusive are accessible in aqueous solution.

4.1 Sources of manganese

The commercially valuable ores of manganese are oxides or hydrated oxides. The most important is pyrolusite, MnO_2. Nearly 50 per cent of land-based manganese reserves occur in the Republic of South Africa, and nearly 40 per cent in the USSR. There are no reserves of commercial value in the United States. In the past 30 years, it has become clear that vast quantities of manganese occur on the deep ocean floor in the form of nodules with a manganese content of 15–30 per cent. If the necessary technology can be developed, the United States is especially anxious to rake some in!

In 1984, approximately 9 million tonnes of manganese was consumed, mostly as ferromanganese, an alloy used in the steel industry. It contains about 80 per cent manganese and 20 per cent iron.

4.2 Manganese metal

In Table 5, data on manganese metal and some of its compounds are compared with the corresponding values for iron and cobalt. These data suggest that in metallic manganese, the atoms are less strongly bonded together than they are in iron. Thus, the melting temperature is nearly 300 °C lower, and the value of $\Delta H_f^\ominus(M, g)$ is nearly 150 kJ mol^{-1} less.

Table 5 Some properties of the elements manganese, iron and cobalt

Element	Melting temperature °C	$\dfrac{\Delta H_f^\ominus(M, g)}{kJ\,mol^{-1}}$	$\dfrac{\Delta G_f^\ominus(MO, s)}{kJ\,mol^{-1}}$	$\dfrac{\Delta G_f^\ominus(M^{2+}, aq)}{kJ\,mol^{-1}}$	$\dfrac{\Delta G_f^\ominus(MS, s)}{kJ\,mol^{-1}}$
Mn	1 244	281	−363	−228	−218
Fe	1 535	416	−251	−90	−100
Co	1 495	425	−214	−54	−80

Both kinetically and thermodynamically, manganese is more easily oxidised than iron, especially if slightly impure. In air, surface oxidation occurs, but the metal burns if it is finely divided. Manganese dissolves more rapidly than iron in dilute acids:

$$Mn(s) + 2H^+(aq) = Mn^{2+}(aq) + H_2(g) \qquad 19$$

and some hydrogen is evolved even in water.

The fact that oxidation to the +2 state is thermodynamically more favourable for manganese than for iron is apparent from the values of $\Delta G_f^\ominus$ for the oxides, sulphides and aqueous ions in Table 5. Thus, metallic manganese reduces a solution of $Fe^{2+}(aq)$ to iron:

$$Mn(s) + Fe^{2+}(aq) = Mn^{2+}(aq) + Fe(s) \qquad 20$$

☐ Use the data in Table 5 to find $\Delta G_m^\ominus$ for reaction 20.

SLC 7 ■ $\Delta G_m^\ominus = -138$ kJ mol^{-1}; following the procedure in the Second Level Course,

$$\Delta G_m^\ominus = \Delta G_f^\ominus(\text{Mn}^{2+}, \text{aq}) - \Delta G_f^\ominus(\text{Fe}^{2+}, \text{aq})$$

As you will see, the easier oxidation of manganese lies behind the contribution that the element makes to the steel industry.

4.2.1 Preparation of metallic manganese

Pure manganese metal is made electrolytically. When pyrolusite is roasted, it decomposes to lower oxides, which can then be dissolved in sulphuric acid; for example:

$$\text{MnO(s)} + 2\text{H}^+(\text{aq}) = \text{Mn}^{2+}(\text{aq}) + \text{H}_2\text{O(l)} \qquad 21$$

The solution is neutralised with ammonia to a pH of 6–7, and then electrolysed with a lead anode and a stainless steel cathode. Manganese metal is deposited at the cathode.

The element is consumed largely as the alloy, ferromanganese, which is made by reducing the appropriate mix of manganese oxide and iron oxide ores with coke in an electric furnace:

$$\text{MnO}_2(\text{s}) + 2\text{C(s)} = \text{Mn(l)} + 2\text{CO(g)} \qquad 22$$

$$\text{Fe}_2\text{O}_3(\text{s}) + 3\text{C(s)} = 2\text{Fe(l)} + 3\text{CO(g)} \qquad 23$$

The liquid alloy solidifies on cooling.

4.2.2 Manganese and the steel industry

Iron that is obtained directly from a blast furnace is called pig iron, and a typical analysis is shown in Table 6 as Stage A. The material is brittle, principally because the carbon content is too high. When this is lowered to about 0.4 per cent, a steel is formed, a tough ductile metal, which can be cheaply shaped by mechanical working.

Table 6 Percentage composition of a sample of Victorian steel at three stages of manufacture: A, as tapped from the blast furnace; B, after aeration in a Bessemer converter; C, final composition after treatment with ferromanganese alloy

Stage	C	Si	P	S	Mn	Fe
A	3.10	0.98	0.10	0.06	0.40	95.0
B	0.04	0.02	0.10	0.06	0.01	99.6
C	0.45	0.04	0.10	0.06	1.15	98.0

Today, annual world production of steel is about 700 million tonnes. The modern steel age was initiated by the experiments of Henry Bessemer in a St Pancras workshop in 1855. They led to the Bessemer converter, a rotatable container in which air could be blown through molten pig iron at about 1 200 °C (Figure 18). To begin with, the converter was loaded with molten pig iron in the horizontal position. The air blast was then turned on and the vessel rotated to the vertical. At first, the air began oxidising the silicon and manganese in the steel and a short flame issued from the mouth of the converter:

$$\text{Mn(melt)} + \tfrac{1}{2}\text{O}_2(\text{g}) = \text{MnO(s)} \qquad 24$$

$$\text{Si(melt)} + \text{O}_2(\text{g}) = \text{SiO}_2(\text{s}) \qquad 25$$

The heat evolved in these reactions actually *raised* the temperature of the molten iron to about 1 600 °C, a temperature at which carbon was oxidised to carbon monoxide. The gas burnt at the converter's mouth, lengthening the flame, and turning it a bright orange yellow, edged with blue and shot through by showers of brilliant sparks. When the carbon in the charge had been consumed, the flame shortened. This was the signal to swing the converter back to the horizontal position and to shut off the air blast. On the surface of the molten iron was a slag consisting of the oxidised manganese and silicon, together with some oxidised iron. This was skimmed off, and the molten metal could then be poured and cast into ingot moulds.

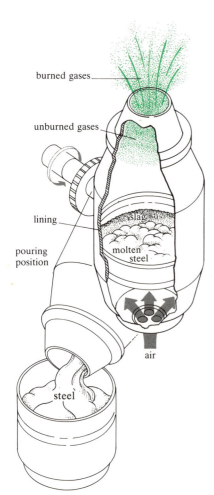

Figure 18 The Bessemer converter.

The ingots so obtained are denoted as Stage B in Table 6.

☐ Is the carbon content satisfactory?

■ No; it has fallen too much, and is now below the desired level of 0.4 per cent.

This snag, the excessive consumption of carbon by the air blast, was the first of three problems that had to be solved before the **Bessemer process** could become a commercial success. The second was created by oxygen from the air blast that remained in the molten steel; as the metal cooled, it combined with some residual carbon to form bubbles of carbon monoxide gas, which caused the metal to froth up in its mould. The third was the percolation of the metal by iron(II) sulphide films, formed from the residual sulphur in the steel; when attempts were made to forge the steel, the films melted and the grains of iron fell apart.

All three problems were solved in 1857 by Robert Mushet, a metallurgist from the Forest of Dean. He added an alloy of manganese, iron and carbon to the converter immediately after the air blast had been cut off. The carbon solved the first problem by restoring the carbon content to the desired level (see Stage C of Table 6), and the manganese solved the second and third problems. It combined with dissolved oxygen via reaction 24, and in its presence, the sulphur in the steel ended up as harmless globules of MnS, rather than as films of FeS. Mushet later remarked, 'I saw then, that the Bessemer process had been perfected, and that with fair play, untold wealth would reward Mr Bessemer and myself'.

Within 10 years, British steel was less than a quarter of its 1856 price. But the biggest developments occurred in the United States. There, from 1872 onwards, a Scottish immigrant, Andrew Carnegie, built a new business concept of **vertical integration** around the Bessemer process (Figure 19). Besides blast furnaces and steel mills in Pittsburgh, he gained control of low-phosphorus iron ores near Lake Superior, fleets of ore ships on the Great Lakes, a railroad from Lake Erie to Pittsburgh, coal mines and coking ovens in Pennsylvania, and factories for producing finished products such as rails and wire. In 1901, a syndicate of bankers headed by J. P. Morgan bought him out for 250 million dollars. By combining Carnegie's business with its own holdings, the syndicate created the US Steel Corporation, which was then easily the world's largest industrial combine.

The Bessemer process dominated world steelmaking until well into the 20th century. Today other metals as well as manganese are used as oxygen and sulphur scavengers in the modern *Basic Oxygen* steelmaking process. Here the most striking difference from Bessemer's method is the use of pure oxygen rather than air as the oxidising agent. This avoids any incorporation of nitrogen into the steel.

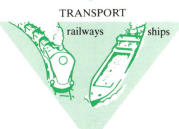

Figure 19 Andrew Carnegie's steel corporation was a pioneering example of vertical integration built around the Bessemer process. His organisation controlled all stages of production from the mining of raw materials to the marketing of finished products.

SAQ 13 Robert Mushet discovered that the addition of manganese to the converter eliminated two of the three problems that stood between the Bessemer converter and commercial success. Today we can see that in both cases, the effect depends on a difference between iron and manganese that was discussed in Section 4.2. What difference is this?

SAQ 14 Use the data in Table 5 to calculate $\Delta G_m^\ominus$ for the reaction

$$\text{Mn(s)} + \text{FeS(s)} = \text{MnS(s)} + \text{Fe(s)} \tag{26}$$

What relevance does your value have to your answer to SAQ 13?

4.3 Manganese halides

MnF_2, $MnCl_2$ and $MnBr_2$ can be made by heating manganese metal in a stream of the gaseous hydrogen halide:

$$\text{Mn(s)} + 2\text{HX(g)} = \text{MnX}_2\text{(s)} + \text{H}_2\text{(g)} \tag{27}$$

MnI_2 can be made by heating manganese with iodine in a sealed tube. MnF_2 has the rutile structure (Figure 3). $MnCl_2$ crystallises with the $CdCl_2$ structure, and $MnBr_2$ and MnI_2 with the CdI_2 structure. As you saw at Second Level, the $CdCl_2$

and CdI_2 structures are composed of the layers shown in Figure 14: the only difference between them lies in the way the layers are stacked on top of each other.

All four dihalides have a pinkish colour, and dissolve in water to give pale pink solutions containing the $Mn^{2+}(aq)$ ion.

No other chlorides, bromides or iodides of manganese can be prepared at room temperature. At normal temperatures, the only other known halides are the higher fluorides MnF_3 and MnF_4. MnF_3 is a red-purple solid made by heating MnF_2 in fluorine:

$$MnF_2(s) + \tfrac{1}{2}F_2(g) \xrightarrow{250\,°C} MnF_3(s) \qquad 28$$

At 550 °C, solid MnF_3 will react with more fluorine to form gaseous MnF_4. This condenses to give an unstable blue solid, which slowly loses fluorine at room temperature, forming MnF_3.

SAQ 15 When explaining why a compound does not exist, one should always specify the decomposition reaction that makes it unstable. Suggest such a reaction for the compound $MnCl_3$, which does not exist at room temperature.

SAQ 16 List the dihalides and trihalides of manganese that exist at room temperature, and compare them with those of titanium. Write down a general reaction for the decomposition of a trihalide into a dihalide. Does the stability of a trihalide with respect to this reaction increase or decrease from titanium to manganese? Which halogen is best at stabilising the higher of the two oxidation states?

4.4 The oxidation states of manganese in aqueous solution

The relative stabilities of the oxidation states are very different in acid and alkali, so these two cases are examined in turn. When reading Sections 4.4.1 and 4.4.2, **SLC 8** check that you can balance redox equations, a skill that was taught at Second Level. Some SAQs on this topic are included.

4.4.1 Acid solution

Manganese metal dissolves in dilute acids to form very pale pink $Mn^{2+}(aq)$. Further oxidation in acid solution is possible with powerful oxidising agents such as persulphate ion, $S_2O_8^{2-}(aq)$, and sodium bismuthate(V), $NaBiO_3$. Depending on the conditions, the reaction with persulphate can produce any one of three higher oxidation states.

Suppose that *one tiny crystal* of $MnSO_4$ is dissolved in a solution of potassium persulphate, $K_2S_2O_8$, in *dilute* sulphuric acid. Now a drop of aqueous silver nitrate is added; this acts as a catalyst for oxidations by $S_2O_8^{2-}(aq)$. After about 20 minutes, the colourless solution becomes rose red, due to production of $Mn^{3+}(aq)$:

$$2Mn^{2+}(aq) + S_2O_8^{2-}(aq) = 2Mn^{3+}(aq) + 2SO_4^{2-}(aq) \qquad 29$$

However, in dilute acid solution, $Mn^{3+}(aq)$ is unstable, a fact that very quickly becomes apparent if this experiment is repeated with normal, higher concentrations of $Mn^{2+}(aq)$. The tripositive ion is then not observed because it decomposes immediately to give a black precipitate of MnO_2:

$$2Mn^{3+}(aq) + 2H_2O(l) = Mn^{2+}(aq) + MnO_2(s) + 4H^+(aq) \qquad 30$$

☐ What kind of reaction is this?

■ A disproportionation: a single oxidation state (+3) decomposes to a higher one (+4) and a lower one (+2).

The $Mn^{2+}(aq)$ product now rejoins unoxidised $Mn^{2+}(aq)$ to await reoxidation to $Mn^{3+}(aq)$; the end product of the overall reaction is MnO_2:

$$Mn^{2+}(aq) + S_2O_8^{2-}(aq) + 2H_2O(l) = MnO_2(s) + 2SO_4^{2-}(aq) + 4H^+(aq) \qquad 31$$

Finally, a third oxidation product can be obtained by adding solid sodium bismuthate(V), a brown powder, to acid solutions of Mn^{2+}(aq). In about 10 minutes, the liquid above the powder becomes deep purple due to the formation of the permanganate ion, MnO_4^-(aq).

SAQ 17 When $NaBiO_3$ acts as an oxidising agent in acid solution, Bi^{3+}(aq) is produced. Write a balanced equation for the reaction of $NaBiO_3$(s) with Mn^{2+}(aq).

☐ What oxidation states of manganese have been encountered in this Section?

■ 0, +2, +3, +4 and +7 in Mn(s), Mn^{2+}(aq), Mn^{3+}(aq), MnO_2(s) and MnO_4^-(aq), respectively.

The missing oxidation states in the sequence from +2 to +7 are +5 and +6. They cannot be made in acid solution; in alkaline solution, however, they can.

4.4.2 Alkaline solution

We begin with a neutral solution of Mn^{2+}(aq). Addition of OH^-(aq) precipitates a white solid, $Mn(OH)_2$:

$$Mn^{2+}(aq) + 2OH^-(aq) = Mn(OH)_2(s) \qquad 32$$

In the absence of oxygen, this undergoes no change, but in air the precipitate slowly becomes brown:

$$2Mn(OH)_2(s) + \tfrac{1}{2}O_2(g) = 2MnO(OH)(s) + H_2O(l) \qquad 33$$

☐ How do the relative stabilities of manganese(II) and manganese(III) differ in acid and alkali?

■ Manganese(III) is more stable in alkali; air will oxidise $Mn(OH)_2$ to $MnO(OH)$, but not Mn^{2+}(aq) to Mn^{3+}(aq).

Further rapid oxidation to manganese(IV) can be achieved by adding hydrogen peroxide to the alkaline suspension of brown $MnO(OH)$:

$$2MnO(OH)(s) + H_2O_2(aq) = 2MnO_2(s) + 2H_2O(l) \qquad 34$$

Manganese dioxide is the starting material for the preparation of our missing oxidation states, +5 and +6. Manganese(VI) can be obtained in the form of the green manganate ion, MnO_4^{2-}, by fusing MnO_2 with excess KOH either in air, or better with an oxygen-carrying oxidising agent, such as potassium nitrate:

$$MnO_2 + 2KOH + \tfrac{1}{2}O_2 = K_2MnO_4 + H_2O \qquad 35$$

On cooling, a green mass is obtained. It dissolves in water to give a dark green alkaline solution, which, if evaporated in a vacuum, gives dark green crystals of potassium manganate(VI), K_2MnO_4.

If the dark green solution is acidified, the reason why it proved unattainable in acid is immediately apparent: a black suspension of MnO_2 in a purple solution of permanganate is formed:

$$3MnO_4^{2-}(aq) + 4H^+(aq) = 2MnO_4^-(aq) + MnO_2(s) + 2H_2O(l) \qquad 36$$

☐ What kind of reaction is this?

■ Disproportionation: manganese(VI) disproportionates to manganese(IV) and manganese(VII) in acid.

So if equation 36 is regarded as an equilibrium system, equilibrium lies to the left in alkali, and to the right in acid. A common way of making potassium permanganate is to prepare an alkaline solution of MnO_4^{2-} as described above. Disproportionation is then effected by disturbing the equilibrium with an increase in the hydrogen ion concentration, for example by passing CO_2 through the solution to neutralise the alkali.

Finally, the blue manganate(V) ion, MnO_4^{3-}, can be obtained by reducing *very* alkaline solutions of MnO_4^{2-} or MnO_4^- with sodium sulphite, Na_2SO_3. The solution deposits blue crystals of the hydrate, $Na_3MnO_4 \cdot 7H_2O$.

SAQ 18 Write balanced equations for:

(a) the reaction of SO_3^{2-} with MnO_4^{2-} in alkaline solution, which yields SO_4^{2-} and MnO_4^{3-};

(b) the reaction analogous to equation 36 in which acidified MnO_4^{3-}(aq) disproportionates to MnO_2(s) and MnO_4^-(aq).

SAQ 19 In Section 4.4.1, you saw that, at typical test-tube concentrations, Mn^{3+}(aq) is unstable in dilute sulphuric acid. By examining the decomposition equilibrium, suggest how the stability of Mn^{3+}(aq) might be increased.

4.4.3 General comments

Table 7 displays the oxidation states covered in this Section, and Figure 20 summarises some of their more important reactions. Table 7 shows how well manganese illustrates the general characteristic of transition elements that was mentioned in Section 3.6.3: they form a variety of coloured compounds/complexes with oxidation states that frequently differ by just one unit. Thus, for manganese, all oxidation states from $+2$ to $+7$ inclusive can be prepared under alkaline conditions. The interconversion of the oxidation states of manganese in aqueous solution is demonstrated in the first sequence of Videocassette 1.

Table 7 Oxidation states of manganese observed in aqueous solution

Formula	Oxidation state	Appearance
Mn(s)	0	metal
Mn^{2+}(aq)	+2	pale pink
Mn^{3+}(aq)	+3	red
MnO_2(s)	+4	brown–black solid
MnO_4^{3-}(aq)	+5	blue
MnO_4^{2-}(aq)	+6	green
MnO_4^-(aq)	+7	purple

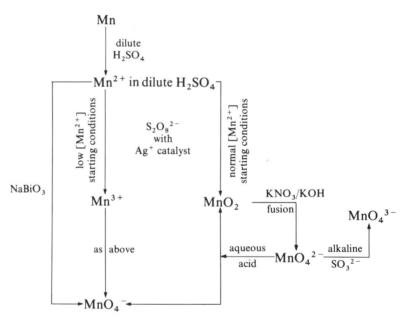

Figure 20 Some reactions of manganese in aqueous solution.

As with titanium chemistry, the colours of the various species have two kinds of origin. Thus, the pale pink of Mn^{2+}(aq), the red of Mn^{3+}(aq) and the blue of MnO_4^{3-} arise from d–d transitions. The green of MnO_4^{2-} and the purple of MnO_4^- are due predominantly to charge-transfer transitions in which an electron residing in an orbital located mainly on oxygen moves to an orbital centred chiefly on manganese.

The ions MnO_4^-, MnO_4^{2-} and MnO_4^{3-} all have regular tetrahedral structures with Mn—O bond lengths in the region of 165 pm. The highest known manganese oxidation state of $+7$ occurs in MnO_4^-, and, as for titanium, is equal to the element's Group number in Mendeléev's Periodic Table.

☐ What is it also equal to?

■ The number of outer electrons in the free manganese atom.

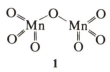

1

The normal oxide of this highest oxidation state is Mn_2O_7. It can be made by adding powdered $KMnO_4$ to concentrated sulphuric acid at $-20\,°C$:

$$2KMnO_4 + 2H_2SO_4 = 2KHSO_4 + Mn_2O_7 + H_2O \qquad 37$$

The oxide is a dangerously explosive green oil, which on standing, slowly loses oxygen and yields MnO_2. It contains discrete molecules with the structure (1) shown in the margin.

SAQ 20 Figure 21 gives the visible and u.v. absorption spectra of aqueous solutions of manganese oxoanions. By referring to these spectra, and to the visible spectrum of Plate 1.1 of *Colour Sheet 1*, explain why the ions have the colours that they do.

4.5 Manganese dioxide and lower oxides

After alloyed manganese metal, manganese dioxide is the most important manganese substance. Its usual crystalline form is pyrolusite, which has the rutile structure (Figure 3) and is grey-black. When prepared from aqueous solutions, MnO_2 often contains some water, and this gives it a brown appearance. This is so whether it is made by oxidising an alkaline suspension of MnO(OH) with peroxide (equation 34) or by oxidising Mn^{2+}(aq) in dilute H_2SO_4 with persulphate (equation 31). The latter reaction can also be performed electrolytically with graphite electrodes at $90\,°C$ in $1\,\text{mol}\,\text{l}^{-1}\,H_2SO_4$, and the solid product can then be stripped off the anode. This is how much of the MnO_2 used in 'dry' batteries (see below) is prepared.

When heated in air, MnO_2 decomposes in two stages; at $600\,°C$, brown Mn_2O_3 is formed, which at $1\,000\,°C$ is converted to black Mn_3O_4, a mixed oxide containing manganese(II) and manganese(III):

$$2MnO_2(s) = Mn_2O_3(s) + \tfrac{1}{2}O_2(g) \qquad 38$$

$$3Mn_2O_3(s) = 2Mn_3O_4(s) + \tfrac{1}{2}O_2(g) \qquad 39$$

If Mn_3O_4, Mn_2O_3 or MnO_2 are heated in hydrogen, grey-green MnO is formed; for example

$$MnO_2(s) + H_2(g) = MnO(s) + H_2O(g) \qquad 40$$

MnO has the NaCl structure, and dissolves in acids to give solutions of manganese(II) salts.

Manganese dioxide has a very low solubility in water and cold dilute acids. It is, however, a powerful oxidising agent, and when heated with moderately concentrated HCl, it dissolves by undergoing reduction and oxidising chloride:

$$MnO_2(s) + 2Cl^-(aq) + 4H^+(aq) = Mn^{2+}(aq) + Cl_2(g) + 2H_2O(l) \qquad 41$$

The dioxide can also be reduced by boiling it with a fairly concentrated solution of H_2SO_4, when oxygen is evolved:

$$MnO_2(s) + 2H^+(aq) = Mn^{2+}(aq) + H_2O(l) + \tfrac{1}{2}O_2(g) \qquad 42$$

For kinetic reasons, sulphuric acid is the only mineral acid in which this reaction goes quickly.

It is the strength of MnO_2 as an oxidising agent that accounts for its main industrial use.

4.5.1 Manganese dioxide and batteries

Manganese dioxide is the oxidising agent in the **Leclanché cell**, the most common kind of 'dry' battery. World-wide, about 10^9 are consumed each year. The reducing agent is zinc metal. A typical example is shown in Figure 22.

Zinc forms the outer casing of the battery, and during discharge of the cell, it is oxidised:

$$Zn(s) = Zn^{2+}(aq) + 2e^- \qquad 43$$

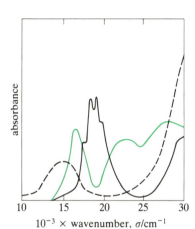

Figure 21 Visible and ultraviolet absorption spectra of manganese oxoanions: full black plot, MnO_4^-(aq); green plot, MnO_4^{2-}(aq); broken black plot, MnO_4^{3-}(aq).

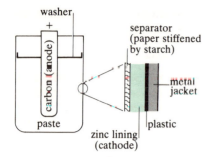

Figure 22 A typical Leclanché cell. The paste comprises solid MnO_2 and carbon; the pores are filled with an aqueous electrolyte ($ZnCl_2$ + NH_4Cl), gelled by starch.

The discharged electrons mean that the zinc casing is the negative pole of the cell. They travel around the external circuit to the positive pole, which consists of a compressed mixture of powdered carbon and manganese dioxide moulded on to a carbon rod. There, the electrons reduce the MnO_2 to $MnO(OH)$, a compound of manganese(III):

$$2MnO_2(s) + 2H_2O(l) + 2e^- = 2MnO(OH)(s) + 2OH^-(aq) \qquad 44$$

The water molecules come from a jelly in which the powdered carbon and MnO_2 are steeped. It consists of an aqueous electrolyte containing dissolved zinc chloride ($ZnCl_2$) and ammonium chloride (NH_4Cl) that has been gelled with starch.

This electrolyte also allows the diffusion of ions that carry current across the gap between the electrodes within the cell. Another contributor to conductivity within the cell itself is the powdered carbon. This has graphite-like conductivity, and helps to transfer electrons from the carbon rod to the MnO_2. Both the carbon and the MnO_2 are also prepared with a porous structure to improve contact with the electrolyte gel.

If the reactions at the two poles describe all the cell's chemistry, then the overall cell reaction can be obtained by adding them together so as to eliminate electrons.

☐ What is the result?

■ Adding equations 43 and 44,

$$Zn(s) + 2MnO_2(s) + 2H_2O(l) = Zn^{2+}(aq) + 2OH^-(aq) + 2MnO(OH)(s) \qquad 45$$

But since $Zn(OH)_2$ is very sparingly soluble, the Zn^{2+}(aq) and OH^-(aq) ions will combine. A better description is therefore

$$Zn(s) + 2MnO_2(s) + 2H_2O(l) = Zn(OH)_2(s) + 2MnO(OH)(s) \qquad 46$$

In fact, there is still uncertainty about the details of the chemistry of the Leclanché cell, and no single cell reaction has gained universal assent. Reaction 46 is favoured by some electrochemists, and has the merit of being fairly simple.

Chemical batteries deliver a substantial voltage when the cell reaction is a redox reaction between a strong oxidising agent and a strong reducing agent. In this case, MnO_2 fulfils the former role, and zinc the latter. Manufacturers usually rate fresh Leclanché cells at 1.5 V. In the so-called alkaline battery, the cell reaction again involves the reduction of MnO_2 by zinc, but no ammonium chloride is present; instead the electrolyte is a gel of concentrated KOH solution.

4.6 Manganese nodules

Thousands of millions of tonnes of manganese occur in the form of ferromanganese nodules on the ocean floors. These were discovered during the famous ocean explorations of HMS Challenger (1872–6). They have the appearance of burnt baked potatoes and their diameter may be as much as 25 cm, although roughly 5 cm is typical. Table 8 shows important metallic constituents of a sample from the Pacific Ocean.

Figure 23 shows a cross-section of a typical manganese nodule. It consists of layers of hydrated oxides, iron being present mainly as iron(III) and manganese as manganese(III) and manganese(IV). In deep-sea nodules these layers seem to be deposited at a rate of 5 mm per million years.

When dissolved in sea-water, manganese is present as Mn^{2+}(aq). As you saw in Section 4.4.1, in acid solution this is rather stable with respect to oxidation reactions such as

$$Mn^{2+}(aq) + \tfrac{1}{2}O_2(g) + H_2O(l) = MnO_2(s) + 2H^+(aq) \qquad 47$$

☐ What feature of the equilibrium in equation 47 suggests that Mn^{2+}(aq) may be less stable in sea-water?

Table 8 Percentages of some metallic elements by mass in a typical manganese nodule from the Northern Pacific Ocean

Metal	Mass per cent
Mn	24.2
Fe	14.0
Si	9.4
Al	2.9
Ni	0.99
Cu	0.53
Co	0.35
Mo	0.05

- Sea-water is not acid; its pH is about 8. This lower hydrogen ion concentration will shift the equilibrium to the right.

The same conclusion is reached through a quantitative study of the thermodynamics of reaction 47 in Block 3: on thermodynamic grounds, manganese is likely to be precipitated from sea-water in the presence of oxygen.

At depths of greater than 10 cm within ocean sediments, there is virtually no oxygen. Here the pore-water contains dissolved manganese as $Mn^{2+}(aq)$. This could have been formed by the action of sea-water on hot submarine volcanic material or, because of the anaerobic conditions, by bacterial reduction of insoluble MnO_2 deposited from the sea-water above. When this $Mn^{2+}(aq)$ migrates upwards through the pore-water, it encounters oxygenated sea-water near the surface of the sediment. The thermodynamic arguments touched on in this Section show that oxidation can then occur, but there remains a kinetic problem: why do manganese and iron accumulate in this very distinctive way? No complete answer to this question exists at present.

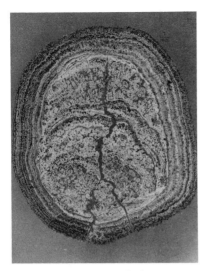

Figure 23 Cross-section of a ferro-manganese nodule, showing the layering.

In current circumstances, the working of a manganese deposit only makes commercial sense if the manganese content is at least 35 per cent. The nodules do not meet this specification but, as Table 8 shows, they contain significant amounts of more valuable metals such as nickel, cobalt and copper. If these were extracted with manganese as a by-product, commercial exploitation of these resources might come to make sense.

4.7 Potassium permanganate

About 40 000 tonnes of this industrially important chemical are produced each year. Most of it is made by first roasting MnO_2 ores with very concentrated KOH solution in air. This yields a green mass of K_2MnO_4 and excess KOH (equation 35), which is then dissolved in aqueous alkali to give $MnO_4^{2-}(aq)$ in KOH of concentration 3 mol l^{-1}. Electrolysis with steel electrodes at 45 °C produces $MnO_4^-(aq)$ at the anode:

$$MnO_4^{2-}(aq) = MnO_4^-(aq) + e^- \qquad \textbf{48}$$

and solid purple $KMnO_4$ eventually crystallises out.

The industrial value of $KMnO_4$ lies in its strength as an oxidising agent. In acid solution it is usually reduced to $Mn^{2+}(aq)$:

$$MnO_4^-(aq) + 8H^+(aq) + 5e^- = Mn^{2+}(aq) + 4H_2O(l) \qquad \textbf{49}$$

This is what happens, for example, when it is used to oxidise acid solutions of $Fe^{2+}(aq)$, $I^-(aq)$ and $H_2O_2(aq)$.

In industry, however, the oxidising properties of permanganate are usually exploited under near-neutral or alkaline conditions, when it is reduced to insoluble MnO_2:

$$MnO_4^-(aq) + 2H_2O(l) + 3e^- = MnO_2(s) + 4OH^-(aq) \qquad \textbf{50}$$

A typical example is the removal of dissolved iron and manganese from public water supplies. The dissolved metals are present as $Fe^{2+}(aq)$ and $Mn^{2+}(aq)$, which give water an unpleasant metallic taste and cause staining. Oxidation with small amounts of dissolved $KMnO_4$ at pH $\approx$ 7 converts them to the insoluble $Fe(OH)_3$ and MnO_2, which can then be filtered off along with the MnO_2 from the reduced permanganate.

Potassium permanganate is also an excellent remover of water tastes and odours caused by organic chemicals. It oxidises these to less offensive or harmless products. Derivatives of phenol, C_6H_5OH, are a particular problem in, for example, power station and other industrial effluents. Potassium permanganate oxidises phenolic compounds to carbon dioxide and water.

4.8 Summary of manganese chemistry

1   The oxidation of manganese metal is both quicker and more thermodynamically favourable than that of iron. The exploitation of this property in the Bessemer process initiated the modern steel industry

2   If we confine ourselves to halides, oxides, aqueous ions and oxoanions, the observed oxidation states are $+2$ to $+7$ inclusive. Of these, $+2$ is especially stable with respect to oxidation or reduction, except in alkaline solution, where it is rather easily oxidised to the $+3$ or $+4$ hydrated oxides. For example, no binary* chlorides, bromides or iodides are known in oxidation states greater than $+2$, and both MnO_2 and MnO_4^- are strong oxidising agents which are fairly easily reduced to manganese(II) in acid solution.

3   The oxidising powers of MnO_2 are exploited in dry batteries, and those of $KMnO_4$ in water treatment and purification.

4   The highest oxidation state of $+7$ is equal to both the Group number in Mendeléev's Table, and to the number of outer electrons in the manganese atom. However, this highest oxidation state is much less stable with respect to reduction than is the $+4$ state of titanium.

Some of the industrially important reactions of manganese and its compounds are summarised in Figure 24. A network of reactions in aqueous solution has already appeared in Figure 20.

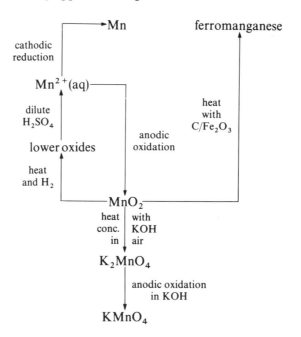

Figure 24 Network of some industrially important reactions of manganese and its compounds.

SAQ 21   When potassium permanganate solution is added to a boiling saturated solution of barium hydroxide, $Ba(OH)_2$, a green barium salt, A, is precipitated. If A is filtered off and stirred with water, a brown suspension, B, in a purple solution, C, is obtained. When B is separated and boiled with 8 mol l^{-1} H_2SO_4, it dissolves to give a pale pink solution, D, with evolution of a gas, E. If D is electrolysed with platinum electrodes, a red coloration, F, immediately appears at the anode. State the chemical formulae that can be associated with each of the substances or colorations A–F.

---

* By binary, we mean compounds containing only two elements; thus a binary chloride contains chlorine and one other element, for example $FeCl_2$.

5 COBALT

Cobalt is the 30th most abundant element (4×10^{-3} per cent by mass) in the Earth's crust and, apart from scandium, is the least abundant of the first-row transition elements. The electronic configuration of the free atom is $[Ar]3d^7 4s^2$. In industry, cobalt metal is by far the most important form of the element. Cobalt blue is the most characteristic glaze coloration in the famous pottery of the Ming dynasty, and cobalt compounds played a central part in the winning of transition-metal chemistry's first Nobel Prize (Section 5.5).

5.1 Sources of cobalt

Two factors have an important influence on the supply of cobalt. The first is that over half of current world production comes from Central Africa, most particularly from Zaïre. The cobalt market is therefore particularly sensitive to signs of political instability in that region. The second is that cobalt is largely a by-product of the mining of other metals, principally copper and nickel. This means that output cannot significantly be increased unless there is a corresponding rise in demand for the primary products.

The cobalt minerals of Zaïre and Zambia are sulphide ores such as carrolite, $CuCo_2S_4$, and compositions based on linnacite, Co_3S_4, in which some of the cobalt in the structure has been replaced by nickel, copper or iron. In 1983, approximately 24 000 tonnes of the element were produced; about 47 per cent of this came from Zaïre, and 13 per cent from Zambia. Most of it was consumed as alloyed cobalt metal.

5.2 Cobalt metal

Table 5 suggests that the binding between atoms in metallic cobalt is stronger than that in manganese, and similar to that in iron. Thus, both the melting temperature and value of $\Delta H_f^\ominus(M, g)$ are close to the values for iron. In general, cobalt is more resistant to common oxidising agents than iron, but the differences are not very great. Lumps of cobalt are not attacked by air below about 300 °C. Above this temperature, the oxide Co_3O_4, which contains both cobalt(II) and cobalt(III), is formed. Like iron, cobalt dissolves in dilute mineral acids, evolving hydrogen and giving pink solutions of the dipositive ion:

$$Co(s) + 2H^+(aq) = Co^{2+}(aq) + H_2(g) \tag{51}$$

The data in Table 5 suggest that cobalt is more thermodynamically stable with respect to oxidation to the +2 oxidation state than either iron or manganese.

☐ Are the data consistent with the fact that iron reduces a solution of $Co^{2+}(aq)$ to the metallic state?

■ Yes; $\Delta G_m^\ominus = -36$ kJ mol^{-1} for the reaction

$$Fe(s) + Co^{2+}(aq) = Fe^{2+}(aq) + Co(s) \tag{52}$$

Large amounts of metallic cobalt are used in materials called **superalloys**. These have high-strength performance, and are especially resistant to corrosion by sulphur during the combustion of sulphur-containing fuels. They are therefore used in jet engines and gas turbines. A typical percentage composition of a superalloy is Co, 55; Cr, 25; Ni, 10; W, 7; Fe, 2; C, 1.

The metals iron, cobalt and nickel all have the familiar magnetic property known technically as *ferromagnetism*. When the three metals are heated, the property is completely lost at a temperature known as the *Curie temperature*. For iron, cobalt and nickel, the Curie temperatures are 770 °C, 1 121 °C and 358 °C, respectively. Cobalt thus has the highest Curie temperature of all the ferromagnetic elements.

5.2.1 Preparation of metallic cobalt

As mentioned in Section 5.1, cobalt is a by-product of the extraction of other metals. Here we describe one method used to process the sulphide copper–cobalt ores of Zaïre. The ore is ground and agitated in water with a frothing agent, which causes the cobalt and copper minerals to adhere to the surface bubbles. They can then be separated from the slag as a sulphide concentrate. This concentrate is roasted in air to convert it to sulphates, which are then dissolved in sulphuric acid. Electrolysis with copper cathodes then extracts the copper, leaving Co^{2+}(aq) in solution. Addition of a solution of $Ca(OH)_2$ precipitates cobalt hydroxides, which are filtered and redissolved in dilute sulphuric acid to give a neutral cobalt(II) solution. Finally, electrolysis deposits cobalt on steel cathodes.

The small but significant cobalt content listed in Table 8 is a major reason for commercial interest in manganese nodules. As noted in Section 4.6, it is the minor ingredients, rather than manganese, which would be the major source of profit.

5.3 Simple aqueous chemistry of cobalt

The pink aqueous ion, Co^{2+}(aq), is extremely stable with respect to oxidation, more so even than Mn^{2+}(aq). Only the most powerful oxidising agents can oxidise it, and the best method is to cool a solution of Co^{2+}(aq) in sulphuric or perchloric acid to 0 °C, and electrolyse with a rotating platinum anode (Figure 25). The pink solution soon turns blue as the ion Co^{3+}(aq) is formed.

☐ Will Co^{3+}(aq) be a strong or a weak oxidising agent?

■ Very strong, because formation from Co^{2+}(aq) is so difficult.

In fact, the blue acid solution of Co^{3+}(aq) is so strong an oxidising agent that it steadily decomposes at room temperature; it reverts to pink Co^{2+}(aq) by oxidising water and evolving oxygen gas:

$$2Co^{3+}(aq) + H_2O(l) = 2Co^{2+}(aq) + 2H^+(aq) + \tfrac{1}{2}O_2(g) \qquad 53$$

Oxidation states of cobalt greater than +3 are not known in aqueous solution.

The addition of sodium hydroxide or sodium carbonate to solutions of Co^{2+}(aq) precipitates blue $Co(OH)_2$ or pink $CoCO_3$, respectively. If these precipitates are filtered off and heated in the absence of air, the olive-green oxide CoO is formed. Like MnO, it has the NaCl structure.

When precipitated by hydroxide ions in the *presence* of air, the hydroxide, $Co(OH)_2$, is oxidised, and gradually changes to the brown cobalt(III) hydroxide:

$$2Co(OH)_2(s) + \tfrac{1}{2}O_2(g) + H_2O(l) = 2Co(OH)_3(s) \qquad 54$$

If filtered off and heated at 150 °C, the hydroxide loses water and forms dark brown CoO(OH).

☐ What is the relationship between equation 53 and equation 54?

■ In equation 53, cobalt(III) is reduced to cobalt(II) and oxidises water to oxygen; in equation 54, oxygen is consumed and oxidises cobalt(II) to cobalt(III).

The interesting point is that in the presence of oxygen and water, the relative stabilities of cobalt(II) and cobalt(III) depend crucially on whether the solution is acid or alkali. In acid, equation 53 operates, and cobalt(II) is formed; in alkali, the relative stabilities are completely reversed: equation 54 is appropriate, and cobalt(III) is produced. We shall look more deeply into this sort of problem in Block 3.

Do the following SAQ now.

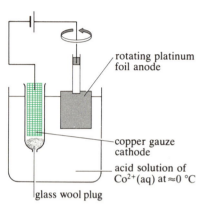

Figure 25 The preparation of Co^{3+}(aq) by electrolytic oxidation at 0 °C at a rotating platinum anode. The bulk of the solution forms the anode compartment. The copper gauze cathode, from which hydrogen gas is evolved, is enclosed by a small tube with a hole at the bottom, which is covered by a glass wool plug. This protects the Co^{3+}(aq) from reduction by the cathode.

SAQ 22 How do these properties of cobalt(II) and cobalt(III) compare with the corresponding ones for manganese(II) and manganese(III)?

These two examples from cobalt and manganese chemistry illustrate the general point that, more often than not, high oxidation states are stabilised in alkaline solution. You will see other examples of this tendency in Section 12.

5.4 Cobalt halides

The known halides of cobalt with their colours and melting temperatures are shown in Table 9. The compounds $CoCl_2$, $CoBr_2$ and CoI_2 can be made by heating cobalt metal with the appropriate halogen; CoF_2 can be obtained by heating $CoCl_2$ in a stream of hydrogen fluoride. The halides $CoCl_2$, $CoBr_2$ and CoI_2 dissolve easily in water to give pink solutions from which pink hexahydrates, $CoX_2.6H_2O$ can be crystallised. The difference in colour between $CoCl_2$ and $CoCl_2.6H_2O$ is exploited by using the anhydrous blue dichloride as an indicator in the desiccant, silica gel: it becomes hydrated and turns pink when the drying capacity has been exhausted.

Table 9 The anhydrous halides of cobalt; melting temperatures are in parentheses

Oxidation state	Fluorides	Chlorides	Bromides	Iodides
+3	CoF_3 pale-brown (decomposes 400 °C)			
+2	CoF_2 pink (1 200 °C)	$CoCl_2$ blue (724 °C)	$CoBr_2$ green (678 °C)	CoI_2 blue–black (515 °C)

The only binary halide containing cobalt in an oxidation state greater than +2 is the light-brown CoF_3, which can be made by heating $CoCl_2$ in fluorine at 250 °C:

$$CoCl_2(s) + \tfrac{3}{2}F_2(g) = CoF_3(s) + Cl_2(g) \qquad 55$$

It is a powerful oxidising agent, which is reduced to cobalt(II) by water, and decomposes when heated above 400 °C:

$$CoF_3(s) = CoF_2(s) + \tfrac{1}{2}F_2(g) \qquad 56$$

SAQ 23 In the range and stability of its halides, which of the elements titanium and manganese does cobalt resemble more? When heated to 700 °C under vacuum, MnF_3 sublimes without decomposition. How does the relative stability of the difluorides and trifluorides change from manganese to cobalt? How does this change fit into the comparison of titanium and manganese which was made in SAQ 16?

5.5 Compounds of cobalt and ammonia

Modern theories of transition-metal chemistry were launched through the study of compounds like those now described. If $CoCl_2.6H_2O$ and ammonium chloride are dissolved in fairly concentrated ammonia, the solution is pink. If air is bubbled through, the colour changes to deep red, and on addition of concentrated HCl, a purple solid with the composition $Co(NH_3)_5Cl_3$ is precipitated. When the experiment is repeated in the presence of a charcoal catalyst, the aeration yields a yellow–brown solution, and the acid precipitates yellow crystals of composition $Co(NH_3)_6Cl_3$. Note that the formulae of the two compounds are identical except that the first contains one less NH_3 group than the second. It is also possible to make a compound in which the number of NH_3 groups is only four: $Co(NH_3)_4Cl_3$.*

When silver nitrate is added to aqueous solutions of these three cobalt compounds, different amounts of chloride are precipitated as AgCl. Every mole of $Co(NH_3)_6Cl_3$ in solution yields 3 moles of AgCl, or 100 per cent of the dissolved chloride; with $Co(NH_3)_5Cl_3$ the yield is 2 moles or 67 per cent, and with $Co(NH_3)_4Cl_3$ it is only 1 mole or 33 per cent. In the period 1880–1900 at least

* In fact, as you will see in Section 5.5.3, it is possible to make two such compounds, but one will do for the moment!

two theories were proposed in explanation of those facts, but the one that has lasted was put forward in 1894 by Alfred Werner (Figure 26), a young Swiss chemist working in Zurich.

Werner argued that there were *two distinct* levels of bonding within the compounds. Firstly, the cobalt exercised a fixed total of binding links in its immediate vicinity, a total that we now call the *coordination number*. For the cobalt–ammonia compounds this total is six, and cobalt is directly linked to six NH_3 groups in a complex unit, which is written $[Co(NH_3)_6]$. Secondly, this complex unit is bound to three exterior chlorines to give a compound $[Co(NH_3)_6]Cl_3$. When the compound is dissolved in water, the three exterior chlorines become aqueous chloride ions, which can be precipitated by silver nitrate. This suggests that the parent compound might be written with the ionic formulation $[Co(NH_3)_6]^{3+}(Cl^-)_3$, and that it can dissociate thus:

$$[Co(NH_3)_6]Cl_3(s) = [Co(NH_3)_6]^{3+}(aq) + 3Cl^-(aq) \qquad 57$$

SLC 9 As noted in a Second Level Course, a unit such as $[Co(NH_3)_6]^{3+}$ is known as a *complex* or *complex ion*. In $[Co(NH_3)_6]Cl_3$, the ammonias are said to occupy the **inner sphere** of the complex; the chlorines are in the **outer sphere**.

On moving to $Co(NH_3)_5Cl_3$, one of the ammonia groups disappears, and if this theory is correct, it has been lost from the inner sphere of the complex $[Co(NH_3)_6]^{3+}$. Werner now argued that, as the coordination number of six must be maintained, one of the exterior chlorines must move into the inner sphere. After these changes, the ionic formulation is $[Co(NH_3)_5Cl]^{2+}(Cl^-)_2$.

□ Does the amount of precipitable chloride obtained from a solution of $Co(NH_3)_5Cl_3$ agree with this formulation?

■ Yes; each mole dissolves to give two moles of $Cl^-(aq)$:

$$[Co(NH_3)_5Cl]Cl_2(s) = [Co(NH_3)_5Cl]^{2+}(aq) + 2Cl^-(aq) \qquad 58$$

SAQ 24 How did Werner write the compound $Co(NH_3)_4Cl_3$? Does the percentage of precipitable chloride implied by your answer agree with the facts?

5.5.1 Testing Werner's theory

To test his theory further, Werner measured the electrical conductivities of solutions of his complexes. These conductivities usually increase with the number of aqueous ions that the complexes yield in solution. Thus, if equations 57 and 58, and the answer to SAQ 24 are correct, $Co(NH_3)_6Cl_3$, $Co(NH_3)_5Cl_3$ and $Co(NH_3)_4Cl_3$ yield four, three and two aqueous ions, respectively. This is corroborated by the electrical conductivities: the value for $Co(NH_3)_6Cl_3$ is the largest of the three, and that for $Co(NH_3)_4Cl_3$ is the least.

Werner made such conductivity measurements, not just for cobalt complexes containing chloride, but for those in which chloride had been partly or wholly replaced by other halides, or by the nitrite group, NO_2^-. Some results are listed in Table 10. They show how the conductivities separate the compounds into well-defined groups that yield 4, 3, 2 or zero ions in solution.

Figure 26 Alfred Werner (1866–1919). Born in Alsace, he later took Swiss nationality and worked in Zurich. In 1913, he was awarded the Nobel Prize for Chemistry for his work on transition-metal chemistry. A sociable man, he liked billiards, cards and drink. The latter probably contributed to an early death from arteriosclerosis.

Table 10 Conductivity values of some cobalt compounds in aqueous solution

Formula	Molar conductivity* $\mathrm{cm^2\,ohm^{-1}\,mol^{-1}}$	Proposed number of aqueous ions
$[Co(NH_3)_6]Cl_3$	431.5	4
$[Co(NH_3)_6](NO_2)_3$	421.9	
$[Co(NH_3)_5Cl]Cl_2$	261.3	3
$[Co(NH_3)_5(NO_2)]Cl_2$	246.4	
$[Co(NH_3)_5(NO_2)](NO_2)_2$	234.4	
$K[Co(NH_3)_2(NO_2)_4]$	99.3	2
$[Co(NH_3)_4(NO_2)_2]Cl$	98.3	
$[Co(NH_3)_3(NO_2)_3]$	8.0†	0

* At a concentration of $0.001\,\mathrm{mol\,l^{-1}}$.

† The expected value for a non-electrolyte is zero, but in practice there is a residual conductivity due to impurities or a reaction with the solvent.

Of particular importance were the results for the compounds $Co(NH_3)_6(NO_2)_3$, $Co(NH_3)_5(NO_2)_3$ and $Co(NH_3)_3(NO_2)_3$. The conductivities suggest that the first two compounds yield four and three ions in solution, respectively.

☐ What does Werner's theory predict for $Co(NH_3)_3(NO_2)_3$?

■ Six-coordination of cobalt demands $[Co(NH_3)_3(NO_2)_3]$; there are no ionisable groups, so the compound should be a *molecular* complex and a non-electrolyte.

This striking prediction meant that the data for $Co(NH_3)_3(NO_2)_3$ were of special importance. In the words of S. M. Jorgensen, Werner's great rival, 'This is the central point in Werner's system; with this it stands or falls'. The last result in Table 10 shows that the conductivity measurements corroborated Werner's predictions.

The adaptability of Werner's theory is illustrated by attempts to obtain from aqueous solutions a cobalt complex with one less NH_3 group than $[Co(NH_3)_3(NO_2)_3]$. Elimination of the NH_3 group is balanced by the appearance of a metallic element as in $KCo(NH_3)_2(NO_2)_4$. As Table 10 shows, the compound is a 1 : 1 electrolyte.

☐ Can Werner's theory account for this?

■ Yes; six-coordination of cobalt accounts for all the NH_3 and NO_2 groups. Writing potassium as K^+, we then have $K^+[Co(NH_3)_2(NO_2)_4]^-$, which is a 1 : 1 electrolyte.

Werner's own famous summary of his experiments is the plot of solution conductivities against the formulae of selected compounds shown in Figure 27. It shows with great clarity the effect of the step-by-step displacement of neutral NH_3 groups by external anionic groups, entering the inner sphere of the cobalt complex.

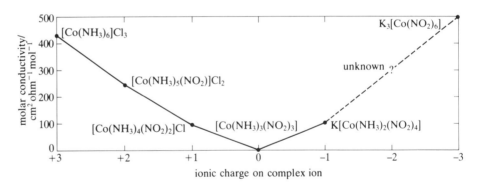

Figure 27 The molar conductivities of a series of cobalt complexes according to Alfred Werner and others.

SAQ 25 Chromium forms a series of six-coordinate complexes similar to those of cobalt. The compound $Cr(NH_3)_4(H_2O)Cl_3$ dissolves in water, and in aqueous solution has a molar conductivity of $261\,cm^2\,ohm^{-1}\,mol^{-1}$. Two-thirds of the total chlorine content can be precipitated from aqueous solution with silver nitrate. Use Werner's theory to write a formula for the compound, and for its dissolution in aqueous solution. Show that your answers are consistent with the conductivity and the precipitable chloride.

SAQ 26 Two compounds with the formula $Co(NH_3)_5Cl(SO_4)$ exist. A solution of one of them gives a precipitate with silver nitrate but not with barium chloride. A solution of the other gives a precipitate with barium chloride but not with silver nitrate. How would you explain these observations?

Table 11 Conductivity values of some platinum compounds in aqueous solution at a concentration of $0.001\,mol\,l^{-1}$

Compound	Molar conductivity $cm^2\,ohm^{-1}\,mol^{-1}$
$Pt(NH_3)_4Cl_2$	261
$Pt(NH_3)_3Cl_2$	116
$Pt(NH_3)_2Cl_2$	2
$KPt(NH_3)Cl_3$	107
K_2PtCl_4	267

SAQ 27 Platinum forms a series of compounds whose formulae and molar conductivities in solution are shown in Table 11. Explain how a common coordination number can be used to rewrite the formulae in a way that explains the values of the molar conductivities.

5.5.2 The anatomy of the complex

One mark of a truly great scientific theory is its fertility: it creates a whole new research programme, generating ideas and experiments that otherwise would not happen. Werner's theory did this, principally because he made bold speculations about the geometry of his complexes. In the case of six-coordinate cobalt, he argued that the surrounding groups were arranged at the corners of an octahedron. In structure **2**, this arrangement is shown for the complex $[Co(NH_3)_6]^{3+}$. Before discussing the momentous consequences of this idea, we shall revise some terminology from the Second Level Course.

SLC 10

A typical complex like that in structure **2** consists of a metallic element surrounded by groups called *ligands*. Some common ligands that you will meet repeatedly in Blocks 1–3 are shown in Table 12. They are listed in states with *closed-shell configurations*. Thus, with the exception of the sulphur in SO_4^{2-}, all the elements in the ligands have noble gas configurations: for example, halogens are listed as halide anions.

Table 12 Some simple ligands

Coordinating atom	Neutral species	Ions
C	CO	CN^-
N	NH_3	NO_2^-*
O	H_2O	O^{2-}, OH^-, SO_4^{2-}, NO_2^-*
halogen		F^-, Cl^-, Br^-, I^-

* NO_2^- can coordinate through nitrogen or oxygen.

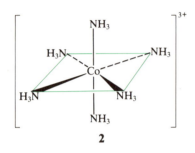

2

The ligands of Table 12 contain electronegative elements such as C, N, O and halogen with lone pairs of electrons, through which the ligand is attached to the central metallic element. For example, in structure **2**, the nitrogen atoms of the NH_3 ligands carry the lone pairs, and it is these atoms that are bound directly to cobalt. The theory of metal–ligand bonding is discussed in Block 2. All we shall do at this stage is to mention *two* very crude but still useful ways of looking at it. Firstly, there is an electrostatic viewpoint: the central metal atom carries a net positive charge, so the ligands orientate themselves with the negative charge of the lone pairs directed towards this positive site; the electrical interaction between them binds the metallic element and ligand together. Secondly, there is a covalent viewpoint: the ligands' lone pairs become electron-pair bonds if the ligands donate them to the metallic element. Compare this with the bond formed when NH_3 donates its lone pair to BCl_3 in the compound $H_3N \to BCl_3$; you met this in a Second Level Course. The comparison shows that this type of bonding need not be confined to metal complexes, and the class of compounds in which it occurs is often given the general name, **coordination compounds**.

SLC 11

Some ligands contain two or more atoms with lone pairs of electrons. The relative positions of these atoms may allow them to form separate bonds to the same metal atom. Some examples are shown in Table 13. They are called **chelating agents**. An example is ethylenediamine, $H_2N-CH_2-CH_2-NH_2$ (en for short), with two such atoms.

☐ Which two?

■ Only the two nitrogen atoms carry lone pairs of electrons, so they are the ones that can bind to the metallic element.

The geometry of an ethylenediamine molecule is such that its two nitrogen atoms can span two *adjacent* corners of the octahedral arrangement taken up by the ligands in cobalt complexes. Thus, three such molecules can complete the octahedron by forming three five-membered rings with a common vertex at cobalt (structure **3**). The resulting complex, $[Co(en)_3]^{3+}$, is usually represented by the simplified diagram of structure **4**.*

Because en has two possible points of attachment through which it can bind to a metal atom, it is known as a **bidentate ligand**. This distinguishes it from **unidentate ligands** such as NH_3, H_2O and halide, which bind through just *one*

---

* When ammonia and its alkyl derivatives (amines), such as ethylenediamine, form complexes with metals, they are called *ammine* ligands.

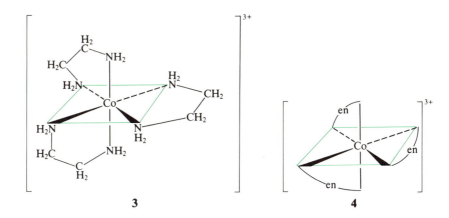

3   **4**

atom. Other bidentate ligands include 1,10-phenanthroline (phen) and 2,2′-bipyridyl (bipy). Table 13 also contains two **polydentate ligands**, which can combine through *more than two* atoms: the tridentate ligand, diethylenetriamine (dien), and the hexadentate ethylenediaminetetraacetate anion (edta^{4-}). The geometry of edta^{4-} is such that the four oxygen and two nitrogen atoms of just one ligand can occupy all six positions around a metallic element in an octahedral complex.

Table 13 Some polydentate (chelating) ligands

Name	Abbreviation	Structure
ethylenediamine	en	$H_2N\text{-}CH_2\text{-}CH_2\text{-}NH_2$
2,2′-bipyridyl	bipy	two pyridine rings joined, N atoms ortho
1,10-phenanthroline	phen	phenanthroline with two N
oxalate	ox^{2-}	$^-O\text{-}C(=O)\text{-}C(=O)\text{-}O^-$
diethylenetriamine	dien	$H_2N\text{-}(CH_2)_2\text{-}NH\text{-}(CH_2)_2\text{-}NH_2$
ethylenediaminetetraacetate	edta^{4-}	$CH_2\text{-}N(CH_2COO^-)_2$ / $CH_2\text{-}N(CH_2COO^-)_2$

Finally notice that the cobalt complexes discussed in Sections 5.5 and 5.5.1 have something more than just their octahedral coordination in common: in each case, the cobalt has the same oxidation state. We define the oxidation state of the metallic element in a complex as the charge left on it when the ligands are removed in their closed-shell configurations.*

This definition ascribes an oxidation state of +3 to cobalt in all the complexes of Sections 5.5 and 5.5.1.

☐ Show that this is the case in the compound $K[Co(NH_3)_2(NO_2)_4]$.

■ From Table 12, the closed-shell configurations of the ligands are NH_3 and NO_2^-; removing $2NH_3$ and $4NO_2^-$ from the anion $[Co(NH_3)_2(NO_2)_4]^-$ leaves Co^{3+}: the anion is a cobalt(III) complex.

SLC 2

* This definition of oxidation state is a natural extension of that used at Second Level, where typical key oxidation states from which others were computed were +1 for hydrogen and the alkali metals, −1 for the halogens in halides and −2 for oxygen in oxides. These are the charges carried by such elements when they are in closed-shell configurations.

The gaseous ion Co^{3+} has the electronic configuration $[Ar]3d^6$, so in the spirit of Section 2.1 it is useful to think of cobalt in cobalt(III) complexes as having a d-electron configuration of $3d^6$. The calculation of oxidation states of transition elements in complexes, and of the associated d-electron configuration is an important skill. SAQ 28 provides you with some practice.

SAQ 28 State the oxidation state of the metallic element, and the associated d-electron configuration in $[Cr(NH_3)Cl(en)_2]Cl_2$ and $[Ni(edta)]^{2-}(aq)$.

Sections 5.5.3 and 5.5.4 deal with the stereochemistry of cobalt complexes. You may find it helpful to follow the arguments if you construct models of the complexes with your model kit.

5.5.3 Geometric isomerism of cobalt(III) complexes

To support his claim that cobalt in cobalt(III) complexes is octahedrally coordinated, Werner drew on an analogy with organic chemistry: the tetrahedral distribution of bonds around carbon generates stereoisomers; would the octahedral distribution around cobalt do the same?

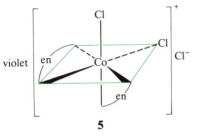

violet

5

In 1889, five years before Werner's theory was published, two compounds with the formula $CoCl_3(en)_2$ had been discovered. One compound was green, and the other was violet. The violet compound can be made by evaporating an aqueous solution of the green one. If it is then treated with hydrochloric acid, the green compound is regenerated. Both compounds have just one precipitable chloride.

To explain the existence of these two compounds, Werner wrote both as $[CoCl_2(en)_2]Cl$, and argued that there was a difference in the distribution of the ligands within the octahedron around cobalt (structures **5** and **6**). In the violet compound, the two chlorides occupied adjacent or *cis*-positions in the octahedron (structure **5**); in the green compound, the two chlorides had opposite or *trans*-positions (structure **6**). Thus, the two compounds were geometric isomers: *cis*-$[CoCl_2(en)_2]Cl$ and *trans*-$[CoCl_2(en)_2]Cl$.

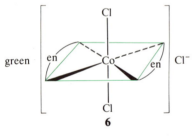

green

6

Now suppose that the two en ligands are replaced by four NH_3 molecules to give the formula $Co(NH_3)_4Cl_3$.

☐ How many compounds do Werner's ideas predict?

■ Again there are two: *cis*- and *trans*-$[Co(NH_3)_4Cl_2]Cl$ (structures **7** and **8**, respectively).

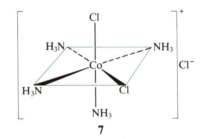 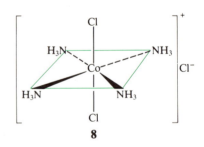

7 **8**

When Werner published his theory, just one compound of formula $[Co(NH_3)_4Cl_2]Cl$ was known, and it was green. But in 1907, after many disappointments, Werner confirmed his prediction by preparing another compound with the same formula. This isomer was violet, and Werner assumed that it was *cis*-$[Co(NH_3)_4Cl_2]Cl$, the counterpart of the green *trans* compound that was already known. Modern X-ray crystallography has entirely confirmed the structures proposed by Werner.

5.5.4 Optical isomerism of cobalt(III) complexes

Another example of the fertility of Werner's theory was that it conjured up the prospect of *optically active* transition-metal complexes. Consider compounds with the formula $[Co(NH_3)Cl(en)_2]Br_2$. The complex cation can exist in both *cis* and *trans* forms. Look first at the *trans* form (structure **9**).

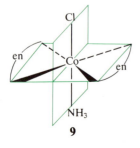

9

□ Does this have any planes of symmetry?

■ Yes; for example the vertical plane shown in structure **9**, which contains Cl, NH₃ and Co.

SFC 5 From the Science Foundation Course, you know that the presence of a plane of symmetry means that the compound is achiral. But in *cis*-[Co(NH₃)Cl(en)₂]⁺ there is no such plane of symmetry: the complex is chiral, and the two non-superimposable mirror images of structures **10** and **11** correspond to two different compounds, which rotate the plane of polarised light in opposite directions. (At this stage of the Course, it is sufficient to take the absence of a plane of symmetry as a mark of chirality.) Thus, in principle, three different stereoisomers with the formula [Co(NH₃)Cl(en)₂]Br₂ exist—the *trans* compound and two optical isomers of the *cis* compound.

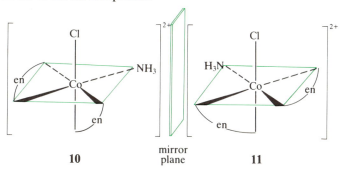

In 1911, Werner and one of his research students, Victor King, succeeded in resolving the *cis* compound into its two distinct chiral forms. This was the first time that optical isomers of a metal complex had been prepared.

SAQ 29 In SAQ 27, you produced four-coordinate formulations for the set of platinum compounds in Table 11 without specifying the geometry of the complexes. Two common kinds of four coordination are tetrahedral and square planar (see Figure 28). The compound [Pt(NH₃)₂Cl₂] exists in two forms, neither of which is optically active. With which of the two types of four-coordination is this observation consistent?

SAQ 30 How many possible stereoisomers exist for the complex [Co(en)₃]³⁺?

SAQ 31 Draw the complete set of stereoisomers for the octahedral complexes [CoBr₂Cl₂(en)]⁻ and [Co(NH₃)₃(H₂O)BrCl]⁺. How many of these stereoisomers are optically active?

SAQ 32 An octahedral complex of cobalt(III) contains just dien and chloride, and its solution gives no precipitate with silver nitrate. What is its formula? If both terminal nitrogens in coordinated dien must occupy positions adjacent to the central nitrogen in the octahedron, how many stereoisomers of the complex should exist? Are any of them optically active?

5.5.5 Werner's work in a wider context

Sections 5.5–5.5.4 were concerned entirely with Werner's work on octahedral cobalt(III) complexes. However, his ideas are obviously applicable to other elements and other geometries. In this Block, you have already met aqueous ions that were written Ti³⁺(aq), Mn²⁺(aq), Co²⁺(aq) and Co³⁺(aq). In general, simple dipositive and tripositive aqueous ions of the first-row transition elements are octahedral complexes of the type [M(H₂O)₆]²⁺(aq) and [M(H₂O)₆]³⁺(aq); representations like Co³⁺(aq) are just shorthand for [Co(H₂O)₆]³⁺(aq).

Again, although octahedral coordination is common to many transition-metal complexes, other types are also plentiful. Tetrahedral coordination in transition-metal complexes is already familiar to you in the oxoanions of manganese (Section 4.4), and some of Werner's most important work was done on square-planar complexes of platinum(II) such as [Pt(NH₃)₄]²⁺, where the four ligands lie at the corners of a square (see SAQ 29). Because his concept of coordination compounds was so productive, you will not be surprised to know that Werner (Figure 26) was awarded the 1913 Nobel Prize for Chemistry.

Figure 28 Two common kinds of four-coordination: (a) square planar; (b) tetrahedral.

5.6 Some cobalt(II) complexes

Section 5.5 described the preparation of $[Co(NH_3)_6]Cl_3$, a cobalt(III) complex, by aerial oxidation of an ammoniacal solution of $CoCl_2$ in the presence of a charcoal catalyst. In the absence of oxygen, no oxidation occurs, and wine-coloured $[Co(NH_3)_6]Cl_2$ is precipitated by the addition of ethanol. $[Co(en)_3]Cl_2$ can be made in a similar way.

$[Co(H_2O)_6]^{2+}$, $[Co(NH_3)_6]^{2+}$ and $[Co(en)_3]^{2+}$ are all octahedral complexes of cobalt(II), with characteristic pink to violet colours. However, cobalt(II) also forms many tetrahedral complexes and these are usually an intense blue. For example, in the presence of dissolved salt, a pink solution of $[Co(H_2O)_6]^{2+}(aq)$ turns blue on boiling, but becomes pink again when the solution is cooled. These colour changes are caused by the shift in equilibrium between an octahedral and a tetrahedral cobalt(II) complex:

$$[Co(H_2O)_6]^{2+}(aq) + 4Cl^-(aq) \rightleftharpoons [CoCl_4]^{2-}(aq) + 6H_2O(l) \qquad 59$$

☐ Is $\Delta H_m^\ominus$ for this reaction positive or negative?

■ Positive: equilibrium lies to the right at 100 °C and to the left at room temperature.

The blue complex can also be generated at room temperature by the addition of concentrated HCl.

In $[CoCl_4]^{2-}$, cobalt is tetrahedrally coordinated to four chlorines; tetrahedral coordination to four oxygens is found in the deep blue complex $[Co(OH)_4]^{2-}(aq)$, which appears when *very concentrated* alkali is added to freshly precipitated $Co(OH)_2$ in the absence of air:

$$Co(OH)_2(s) + 2OH^-(aq) = [Co(OH)_4]^{2-}(aq) \qquad 60$$

Cobalt(II) in a tetrahedral environment is also responsible for the famous blue pottery glazes on Chinese porcelain of the Ming dynasty, ca. 1420 AD (Plate 1.2 of S343 *Colour Sheet 1*). The main body of such objects was made by firing a shaped mixture of kaolin, feldspar and quartz at 1 300–1 400 °C. After cooling, the surface pattern was painted on with a paste made from a cobalt(II) oxide mineral. The whole object was then coated with an aqueous suspension of a low-melting silicate such as feldspar, and the coating vitrified by a second firing. Co^{2+} ions enter into tetrahedral coordination with oxygen atoms in the surface silicate glaze, and on cooling the pattern becomes blue.

Another famous glaze coloration occurs on the green Celadon ware of Northern China, ca. 1100 AD. Particularly fine examples were made during the Sung dynasty (Plate 1.3 of S343 *Colour Sheet 1*). This colour is produced by iron(II), which takes up octahedral coordination with oxygen atoms in the silicate glaze. The colour is similar to, although more intense than, the pale green of the aqueous ion $[Fe(H_2O)_6]^{2+}$. To produce it, the firing procedure had to maintain a reducing atmosphere in the kiln; otherwise oxidation to yellow–brown iron(III) would have occurred. This point is illustrated by Plate 1.4 of S343 *Colour Sheet 1*, which shows two iron-glazed Chinese ornaments of the Western Chin dynasty in 300 AD. The yellow–brown genie and chimera on the left were fired under oxidising conditions; the green of the ram on the right was generated in a reducing atmosphere.

5.7 Higher oxidation states of cobalt

In Sections 3.5 and 4.4.3, you saw that the highest oxidation states of titanium and manganese are equal to the number of outer electrons in the atoms.

☐ On this basis, what should be the highest oxidation state of cobalt?

■ +9; the electronic configuration of the cobalt atom, $[Ar]3d^74s^2$, reveals nine outer electrons.

The highest oxidation state of cobalt that you have met so far in this Block is only +3. However, we have not yet discussed oxide or fluoride complexes in which the highest oxidation states of elements are usually to be found. Thus, if Cs_2CoCl_4 is made by heating CsCl with $CoCl_2$ and then fluorinated at 350 °C, the chlorine in the compound is replaced by fluorine, oxidation occurs, and a complex fluoride of cobalt(IV) is formed:

$$Cs_2CoCl_4(s) + 3F_2(g) = Cs_2CoF_6(s) + 2Cl_2(g) \qquad 61$$

Likewise, when a 3:1 mixture of Cs_2O and Co_2O_3 is heated in oxygen in a sealed tube, black Cs_3CoO_4 is formed. The compound is said to contain cobalt(V), although this has not yet been fully proved. The oxidation state +5 is the highest currently known for cobalt; clearly it falls far short of the number of outer electrons.

5.8 Summary of cobalt chemistry

1 Cobalt metal resembles iron and nickel in being ferromagnetic. Both kinetically and thermodynamically it is somewhat less easily oxidised than iron by air, water and acids. Considerable amounts of cobalt metal are obtained by electrolytic reduction of aqueous solutions of cobalt(II).

2 If we confine ourselves to halides and halide complexes, oxides, aqueous ions and oxoanions, the observed oxidation states are +2 to +5 inclusive. Of these, +4 and +5 are highly unstable to reduction and of little importance. The relative stabilities of the oxidation states +2 and +3 varies greatly with the ligand and the pH. With the simple aqueous ions in acid solution, +3 is very unstable with respect to +2; in alkaline solution, when hydrated oxides are precipitated, the reverse is the case. Oxygen converts cobalt(II) complexes with ammine ligands such as NH_3 and en to the +3 state. The range of oxidation states is much narrower than for manganese, and the highest known falls well short of the number of outer electrons.

3 Cobalt(III) complexes are very numerous and usually octahedral. By studying their properties, Alfred Werner established the concept of a coordination compound, and prepared stereoisomers of cobalt(III) ammine derivatives.

4 Cobalt(II) forms both octahedral and tetrahedral complexes; the latter are often an intense blue.

Some important reactions of cobalt and its compounds are summarised in Figure 29. The preparation of the aqueous ion $Co^{3+}(aq)$, and of the complex $[Co(NH_3)_6]Cl_3$ are demonstrated in the first sequence of Videocassette 1.

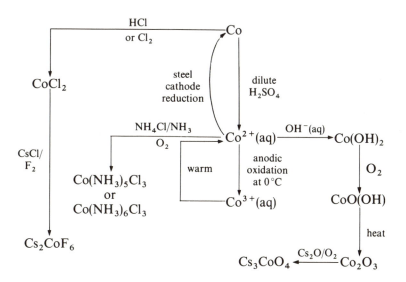

Figure 29 Some reactions of cobalt and its compounds.

We recommend that you now view the first sequence of Videocassette 1, referring to the S343 Audiovision Booklet for guidance.

6 WIDENING THE PERSPECTIVE

As you progressed through the chemistry of titanium, manganese and cobalt in Sections 3, 5, 8, there emerged the possibility of generalisations about the entire transition series. For example, with titanium, near the beginning of the series, the +2 oxidation state in halides and aqueous ions is easily oxidised to +3; conversely with cobalt near the end, the +3 oxidation state is easily reduced to +2. Does the stability of the +2 oxidation state increase steadily across the first transition series? Likewise, you saw that for titanium and manganese, the highest-known oxidation state was equal to the number of outer electrons in the free atom. However, for cobalt, it fell well below this number. For how many other elements in the first transition series does this generalisation fail?

From now on, we shall leave the element-by-element standpoint of earlier Sections, and take a bird's-eye view of the entire first transition series. A new and important aspect will be the injection of quantitative physical properties such as thermodynamic data into the argument. Again, we shall be concerned largely with simple binary compounds, aqueous ions and oxoanions. Not only will this study answer the two questions that have just been posed; it will also reveal the general drift in other important properties within the first transition series. We begin with the elemental state of the first-row transition elements.

7 THE ELEMENTAL STATE

The fact that all transition elements are metals is just one illustration of the fact that the transition elements are more alike than are the typical elements: the latter include both metals and non-metals. Apart from manganese, the metals of the first transition series all have one of the three common metallic structures, the body-centred cubic (bcc), the hexagonal close-packed (hcp) or the cubic close-packed (ccp) at room temperature. These three structures were described in a Second Level Course. The structures of the first-row transition metals at room temperature are given in Table 14.

SLC 12

Notice that chromium and iron have the same structure as sodium, cobalt is isostructural with magnesium, and copper is isostructural with aluminium. Evidently, the introduction of d electrons does not greatly influence the structure of metals. However, it does influence other important properties such as the melting temperatures and boiling temperatures. The former are also recorded in Table 14. For comparison, the melting temperatures of sodium, magnesium and aluminium are 98 °C, 651 °C and 660 °C, respectively. You can see that with the exception of zinc, all the first-row transition metals melt at least 400 °C higher than aluminium. The transition metals are also harder than the typical metals. We can postulate that if the transition elements are harder and have higher melting temperatures and boiling temperatures than the typical elements, then this must be attributed to the stronger forces that bind their atoms together. These can be crudely correlated with electronic configuration. In Figure 30 we have plotted the melting temperatures of metals from potassium to zinc. Examine this plot.

Table 14 Structures of the first-row transition elements at room temperature and their melting temperatures

Element	Structure	m.t./°C
Sc	ccp	1 539
Ti	hcp	1 660
V	bcc	1 890
Cr	bcc	1 857
Mn	*	1 244
Fe	bcc	1 535
Co	hcp	1 495
Ni	ccp	1 453
Cu	ccp	1 083
Zn	hcp	420

* The structure of manganese is complex, but the coordination number of 12 is as large as in the common metal structures.

□ What four elements have the lowest melting temperatures?

■ Potassium, calcium, copper and zinc.

□ What do you know about the electronic configurations of these four atoms?

■ The 3d shell is either empty or full. They have no unpaired d electrons.

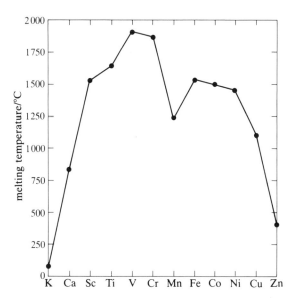

Figure 30 Melting temperatures of the elements from potassium to zinc inclusive.

We can think of unpaired d electrons in the free atoms as electrons available for the formation of particularly strong electron-pair bonds. Notice that between calcium and scandium, where a d electron first appears, there is a jump of nearly 700 °C in the melting temperature. The presence of one or more unpaired d electrons thus leads to higher interatomic forces, and therefore to high melting temperatures and boiling temperatures. However, we cannot make this simple argument more precise by saying that the melting temperatures increase with the number of unpaired electrons in the ground-state configuration of the atom: manganese, for example, with five unpaired d electrons, has a lower melting temperature than titanium or nickel, which have only two. Nevertheless, even the most sophisticated theories suggest that strong bonds formed by d electrons are responsible for the particularly high melting and boiling temperatures of the transition metals.

Most transition metals are not readily attacked by air or water, in contrast to the alkali or alkaline earth metals, and this, together with their hardness and high melting temperatures, gives them great industrial importance. You have seen examples of this in the industrial uses of titanium, manganese, iron and cobalt.

SAQ 33 Elements A and B occur in Period 5 of the Periodic Table and each has four valence electrons. Both are good conductors of electricity. A melts at 1 852 °C and B at 232 °C. Identify A and B.

8 THE OXIDATION STATE +2

8.1 Aqueous ions

When a sample of one of the five metals, chromium, manganese, iron, cobalt or nickel is dropped into dilute sulphuric acid, a reaction occurs. The reactions occur at varying speeds, but all five metals dissolve with the evolution of hydrogen gas and formation of dipositive aqueous ions. To ensure that Cr^{2+}(aq) is formed, it is necessary to add the metal to dilute acid in the absence of air because Cr^{2+}(aq) is oxidised by atmospheric oxygen. The formation of dipositive aqueous ions by five consecutive transition metals is a further indication of a degree of similarity in the elements of the first transition series.

Apart from these five, three other first-row transition elements form dipositive aqueous ions; a complete list is shown in Table 15. V^{2+}(aq) and Cu^{2+}(aq) *cannot* be obtained by the action of the metal on dilute hydrochloric or sulphuric acids, but Zn^{2+}(aq) can.

□   Summarise in a single sentence, without naming individual elements, the extent to which the first-row transition metals form dipositive aqueous ions.

Table 15 The known dipositive aqueous ions and their colours for the first-row transition metals

Aqueous ion	Colour
Sc^{2+}	ion not known
Ti^{2+}	ion not known
V^{2+}	lavender
Cr^{2+}	sky-blue
Mn^{2+}	very pale pink
Fe^{2+}	pale green
Co^{2+}	pink
Ni^{2+}	green
Cu^{2+}	blue
Zn^{2+}	colourless

■ The *last eight* of the first-row transition elements form dipositive aqueous ions; only the first two elements, appear not to do so.

SAQ 34 What d-electron configuration do you associate with the only colourless M^{2+}(aq) ion in Table 15? What generalisation from earlier Sections does your answer reinforce?

8.2 Dihalides

Most first-row transition-metal difluorides, dichlorides and dibromides can be obtained by heating the metal in a stream of HF, HCl and HBr gas, respectively. The hydrogen halides are used instead of the halogens because in some cases the halogen will oxidise the dihalide to a higher oxidation state. In the case of the di-iodides, however, iodine is just as satisfactory as HI, because, when heated, HI decomposes to give hydrogen and iodine. These methods work satisfactorily for the dihalides of chromium, manganese, iron, cobalt and nickel; for example

$$\text{Fe(s)} + 2\text{HCl(g)} \xrightarrow{500\,°C} \text{FeCl}_2(\text{s}) + \text{H}_2(\text{g}) \qquad 62$$

$$\text{Co(s)} + \text{I}_2(\text{g}) \xrightarrow{500\,°C} \text{CoI}_2(\text{s}) \qquad 63$$

Besides these dry methods, it is also possible to obtain dihalides from aqueous solution, the initial halide obtained being a hydrated salt. Thus manganese, iron and cobalt dissolve in hydrochloric acid to give solutions containing chloride and dipositive metal ions. By concentrating and then chilling the solutions, crystals of the hydrated salts $MnCl_2.4H_2O$, $FeCl_2.4H_2O$ and $CoCl_2.6H_2O$ are obtained. The water of crystallisation can be removed by warming these hydrates with thionyl chloride, $SOCl_2$. Thionyl chloride is a liquid that fumes in moist air because of the hydrolysis reaction:

$$\text{SOCl}_2(\text{l}) + \text{H}_2\text{O(g)} = \text{SO}_2(\text{g}) + 2\text{HCl(g)} \qquad 64$$

□ What happens when a solid hydrated halide is refluxed with liquid thionyl chloride?

■ Thionyl chloride reacts with the water of crystallisation, the gaseous products, SO_2 and HCl, vaporise, and the anhydrous dihalide is left.

Finally, as we shall see later, some first-row transition elements that form dihalides readily form higher halides too. In such cases, the dihalide can usually be obtained by heating the higher halide with a suitable reducing agent such as hydrogen, or the metal. This is true of chromium; for example

$$\text{Cr(s)} + 2\text{CrF}_3(\text{s}) \xrightarrow{1000\,°C} 3\text{CrF}_2(\text{s}) \qquad 65$$

$$\text{CrCl}_3(\text{s}) + \tfrac{1}{2}\text{H}_2(\text{g}) \xrightarrow{500\,°C} \text{CrCl}_2(\text{s}) + \text{HCl(g)} \qquad 66$$

The range of known dihalides is shown in Table 16. Examine this Table.

Table 16 The range of known dihalides of the first-row transition elements

Sc	Ti	V	Cr	Mn	Fe	Co	Ni	Cu	Zn
none known	all known except TiF_2	all known	all known	all known	all known	all known	all known	all known except CuI_2	all known

□ Summarise in a single sentence the extent to which the first-row transition metals form dihalides.

■ All dihalides of the first-row transition metals are known except the dihalides of scandium, TiF_2 and CuI_2.

□ Do you detect any relationship between this generalisation, and the generalisation made earlier about the occurrence of dipositive aqueous ions?

■ In the case of the dipositive aqueous ions, the elements for which such ions are unknown are scandium and titanium, which are to be found at the beginning of the transition series. Five of the six unknown dihalides are also compounds of these two elements.

Attempts to make these dihalides and dipositive ions have certainly been made, so they may well be unknown because they are unstable with respect to some decomposition reaction. We explore possible decomposition reactions in later Sections.

8.2.1 Structures of dihalides

With an important qualification mentioned later, the structures of the dihalides of the first-row transition metals can be described in two short statements:

(i) All difluorides have the rutile structure (Figure 3).

(ii) All the dichlorides, dibromides and di-iodides except those of zinc, have $CdCl_2$ or CdI_2 layer-type structures. The three-deck layer found in both structures was shown in Figure 14.

□ What generalisation can you make about the coordination of the metal in transition-metal dihalides?

■ Both in rutile, and in the $CdCl_2$ or CdI_2 layer structures, the metal is six-coordinated by halogen atoms. These six halogens lie at the corners of an octahedron.

This octahedral coordination of the metallic element occurs in all first-row transition-metal dihalides except $ZnCl_2$, $ZnBr_2$ and ZnI_2, in which the coordination is tetrahedral.

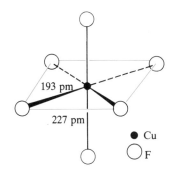

Figure 31 The distorted octahedral coordination around copper in CuF_2.

Now for the qualification mentioned at the beginning of this Section. In the chromium dihalides and the copper dihalides the octahedron of halogens around the metal is strongly distorted as if two opposite corners of the octahedron had been pulled apart. We show the internuclear distances for the octahedral coordination of copper difluoride in Figure 31. Such large distortions from regular octahedral coordination mean that chromium and copper difluoride have similarly highly distorted forms of the rutile structure, whereas the other known dihalides of chromium and copper have highly distorted forms of the CdX_2 layer structures.

□ What d-electron configuration do you associate with chromium and copper in their dihalides?

■ $3d^4$ and $3d^9$, respectively—the d-electron configurations of the gaseous M^{2+} ions.

As you will see in Block 2, there are special reasons why d^4 and d^9 systems should exist in distorted octahedral coordination. However, even when account is taken of the distortions, strong structural similarities obviously exist between the dihalides formed between the first-row transition metals and a particular halogen.

SAQ 35 Describe the coordination around the metal and the fluorine in NiF_2, and the metal and the bromine in $FeBr_2$.

8.2.2 Radii of dipositive ions

How do the ionic radii of the dipositive ions vary across the first transition series? Consider the electronic configurations of the dipositive ions in the series that runs from Ca^{2+} to Zn^{2+}. In Section 2.1 you saw that Ca^{2+} has the electronic configuration of argon, $[Ar]3d^0$, that at Zn^{2+} the 3d shell is full and the configuration is $[Ar]3d^{10}$, and that from Ca^{2+} to Zn^{2+} the electronic configuration of the dipositive ions changes uniformly from $[Ar]3d^0$ to $[Ar]3d^{10}$. Thus, as we move from one element to the next in the series Ca^{2+} to Zn^{2+}, one proton is added to the nucleus and one electron is added to the 3d shell. Let us consider these two effects in turn.

□ First, what effect does successive addition of protons tend to produce in the sizes of M^{2+} ions?

■ The addition of a proton leads to an increase in the nuclear charge, which, by itself, tends to pull in the outer electrons and to decrease the size of the ion.

□ Now consider the successive addition of d electrons. What change does this tend to produce in the sizes of M^{2+} ions?

■ The addition of a 3d electron tends to increase the repulsive forces between the outer electrons and to lead to an increase in the size of the ion.

SLC 13

Either of these effects might predominate, so the radii might increase or decrease in the series from Ca^{2+} to Zn^{2+}. But perhaps we can take a hint from the covalent radii of the elements from lithium to fluorine, which were discussed in a Second Level Course. Covalent radii obtained from experimental internuclear distances decrease steadily from lithium to fluorine as protons are added to the nuclei, and electrons are added to the 2s and 2p levels; clearly, the increasing nuclear charge predominates over the increasing electron repulsion in this case. If the increasing nuclear charge shows the same predominance in the series Ca^{2+} to Zn^{2+}, a smooth decrease in radius should again be observed.

These arguments allow us to make an informed guess at the variation in radius: the radii of the M^{2+} ions should decrease smoothly, but not very substantially, across the series. Let us see how this expectation is borne out. We can obtain values for the radii of the dipositive ions from the metal–fluorine distances in difluorides, and from the radius of the fluoride ion. The average metal–fluorine distances in the rutile structures of some transition-metal difluorides have been obtained by X-ray crystallography and are recorded in Table 17. The usual radius assigned to the fluoride ion is 133 pm.

Table 17 Internuclear distances in difluorides

Metal difluoride	M–F distance/pm	$r(M^{2+})$/pm
VF_2	209	
MnF_2	212	
FeF_2	208	
CoF_2	205	
NiF_2	200	
ZnF_2	204	

□ Calculate values for the radii of the ions V^{2+}, Mn^{2+}, Fe^{2+}, Co^{2+}, Ni^{2+} and Zn^{2+}, and record them in Table 17.

■ You should have subtracted 133 pm from the metal–fluorine distance to obtain the radius of M^{2+}.

Plot the radii of V^{2+}, Mn^{2+}, Fe^{2+}, Co^{2+}, Ni^{2+} and Zn^{2+} obtained in this way in Figure 32. The radius of Ca^{2+} is 99 pm and has already been recorded in Figure 32 for comparison. Now draw straight lines between the calcium and vanadium points, between the vanadium and manganese points, between the manganese and iron points, and so on across the series. You now have a plot of the variation in the ionic radii of the dipositive ions.

The ionic radii of the dipositive transition-metal ions lie in the narrow range 73 ± 6 pm, so it is not surprising that a particular set of dihalides have similar, octahedrally coordinated structures. The value for Ca^{2+} at 99 pm is considerably greater. According to theories of ionic structures, this could lead to higher coordination numbers in calcium compounds, and this sometimes happens. CaF_2, for example, has the eight-coordinate fluorite structure.

□ Is the *overall* trend in the radius for it to decrease as we move from Ca^{2+} to Zn^{2+}?

■ Yes. This supports our belief that the increasing nuclear charge dominates the radius variation.

However, the decrease in ionic radius is not a regular one. If a smooth curve is drawn through the calcium, manganese and zinc points, where the ions have the configurations d^0, d^5 and d^{10}, the other points lie below this curve by amounts that are a maximum at vanadium and nickel, where the ions have the configurations d^3 and d^8, respectively. These irregularities will be explained in Block 2. Here, our main concern is with the point that the structural similarities between transition-metal dihalides are matched by expected similarities in the radii of the dipositive ions.

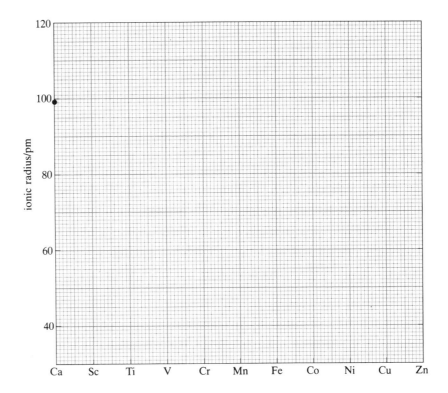

Figure 32 Ionic radii of the dipositive ions of the first transition series (to be completed by students).

SAQ 36 If ScF_2 and TiF_2 could be prepared, what structure would you expect them to have?

8.3 Some reactions within the dipositive oxidation state

In this Block we are concentrating on the chemistry of aqueous ions, oxides, halides and simple salts. In reactions in which these compounds or ions are interconverted and in which there is no change in oxidation state, the first-row transition metals behave similarly. Thus, the eight dichlorides of the elements vanadium to zinc dissolve in de-aerated water under nitrogen to form dipositive ions:

$$MCl_2(s) = M^{2+}(aq) + 2Cl^-(aq) \qquad 67$$

Likewise, addition of sodium hydroxide or sodium carbonate to solutions of these ions in most cases precipitates insoluble hydroxides or carbonates, as you saw for manganese and cobalt in Sections 4 and 5.

$$M^{2+}(aq) + 2OH^-(aq) = M(OH)_2(s) \qquad 68$$

$$M^{2+}(aq) + CO_3^{2-}(aq) = MCO_3(s) \qquad 69$$

Similarities of this kind are not maintained when we turn to redox reactions. As we shall now find, the behaviour of the transition elements in redox reactions is a colourful property, which reveals great variety between the different metals.

9 THE OXIDATION STATE +3

9.1 Aqueous ions

Table 18 tells you the full range of known tripositive aqueous ions of the first row transition metals.

Table 18 The known tripositive aqueous ions and their colours for the first-row transition elements

Aqueous ion	Colour
Sc^{3+}	colourless
Ti^{3+}	purple
V^{3+}	green
Cr^{3+}	green–violet
Mn^{3+}	claret
Fe^{3+}	yellow–brown*
Co^{3+}	blue
Ni^{3+}	ion not known
Cu^{3+}	ion not known
Zn^{3+}	ion not known

* Strictly, pale violet; for a comment on the colour of both this and Cr^{3+}(aq), see the S343 *Audiovision Booklet*.

☐ Summarise concisely, the extent to which the first-row transition metals form tripositive aqueous ions.

■ The first seven elements of the transition series form tripositive aqueous ions: the tripositive ions of the last three members, Ni^{3+}(aq), Cu^{3+}(aq) and Zn^{3+}(aq), are unknown.

☐ In what way does this statement contrast with the statement describing the range of known *dipositive* aqueous ions?

■ The missing tripositive ions occur at the end of the transition series but the unknown dipositive ions, Sc^{2+}(aq) and Ti^{2+}(aq), occur at the beginning.

SAQ 37 What change in the *relative* stabilities of dipositive and tripositive aqueous ions does this suggest from scandium to zinc?

We shall now look for further evidence of the hypothesis advanced by SAQ 37 in the chemistry of chromium, manganese, iron, cobalt and nickel. We shall draw conclusions about the relative stabilities of M^{2+}(aq) and M^{3+}(aq) from simple test-tube observations. We shall not even try to concentrate on thermodynamic stability: that stage will come later. The whole series of experiments is shown in the second sequence of Videocassette 1.

Imagine that we have solutions of Cr^{2+}(aq), Mn^{2+}(aq), Fe^{2+}(aq), Co^{2+}(aq) and Ni^{2+}(aq) in dilute sulphuric acid. The solutions have been prepared not in air, but in an atmosphere of nitrogen and stoppered so that they contain no oxygen. The situation is shown in Figure 33a.

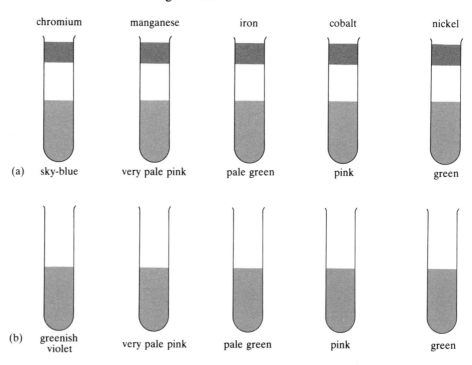

Figure 33 Solutions of metals in dilute sulphuric acid: (a) M^{2+}(aq) prepared by dissolution of the metals chromium to nickel inclusive under nitrogen gas; (b) the effect of admitting air to the solutions in (a).

Imagine now that the stoppers are removed, thus admitting oxygen. With Mn^{2+}, Fe^{2+}, Co^{2+} and Ni^{2+}, no colour change is observed. However, the sky-blue colour of the solution of Cr^{2+}(aq) slowly becomes a greenish-violet. This green–violet colour is due to the growing presence of Cr^{3+}(aq), formed by oxidation of Cr^{2+}(aq) by oxygen and aqueous hydrogen ions:

$$4Cr^{2+}(aq) + O_2(g) + 4H^+(aq) = 4Cr^{3+}(aq) + 2H_2O(l) \qquad 70$$

The situation is shown in Figure 33b.

- □ Is this special behaviour of $Cr^{2+}(aq)$ in accord with our hypothesis?
- ■ Yes. It seems easiest to oxidise $Cr^{2+}(aq)$, the first ion in the series Cr^{2+}, Mn^{2+}, Fe^{2+}, Co^{2+} and Ni^{2+}. On the time-scale of the experiment, no oxidation of $Mn^{2+}(aq)$, $Fe^{2+}(aq)$, $Co^{2+}(aq)$ and $Ni^{2+}(aq)$ by atmospheric oxygen seems to occur.

Now we shall try to oxidise the acid solutions of the aqueous ions $Mn^{2+}(aq)$, $Fe^{2+}(aq)$, $Co^{2+}(aq)$ and $Ni^{2+}(aq)$ with increasingly powerful oxidising agents:

1 Firstly, hydrogen peroxide is added to the solutions of the four aqueous ions. This oxidises $Fe^{2+}(aq)$ to yellow–brown $Fe^{3+}(aq)$, but leaves the other three unaffected. The experiment is shown as the change B → C in Figure 34.

2 The solutions of the three remaining dipositive ions, $Mn^{2+}(aq)$, $Co^{2+}(aq)$ and $Ni^{2+}(aq)$ are then treated with potassium persulphate solution and a trace of $Ag^+(aq)$, which acts as a catalyst. As you know from Section 4.4.1, if the concentration of $Mn^{2+}(aq)$ is small, persulphate oxidises $Mn^{2+}(aq)$ to rose-red $Mn^{3+}(aq)$. However, there is no apparent oxidation of $Ni^{2+}(aq)$ or $Co^{2+}(aq)$ at room temperature. The experiment is shown as the change C → D in Figure 34.

3 $Co^{2+}(aq)$ and $Ni^{2+}(aq)$ are now cooled to 0 °C, and electrolysed at the anode of the cell in Figure 25. As you know from Section 5.3, $Co^{2+}(aq)$ is oxidised to blue $Co^{3+}(aq)$ in this cell. The solution must be chilled because $Co^{3+}(aq)$ is reduced by water:

$$4Co^{3+}(aq) + 2H_2O(l) = 4Co^{2+}(aq) + 4H^+(aq) + O_2(g) \qquad 71$$

Without cooling, this reaction is fast enough to greatly decrease the concentrations of $Co^{3+}(aq)$ which are obtained. Even with this precaution, however, no oxidation of $Ni^{2+}(aq)$ occurs. The experiment is shown as the change D → E in Figure 34.

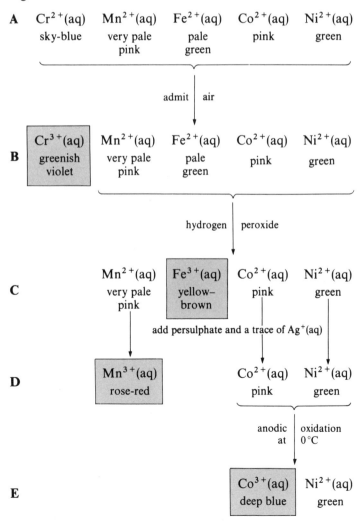

Figure 34 Selective oxidation of $M^{2+}(aq)$ to $M^{3+}(aq)$ by increasingly effective oxidising agents. Oxidation occurs in the sequence chromium, iron, manganese and cobalt.

☐ Does the behaviour of cobalt and nickel support the hypothesis that it gets harder to oxidise M^{2+}(aq) across the series?

■ Yes; Ni^{2+}(aq) is the hardest of the five dipositive ions to oxidise: all attempts to oxidise it have failed. The next most difficult oxidation is that of the dipositive ion of the preceding element, cobalt. Co^{3+}(aq) is produced only in the last and most powerful oxidising stage, D → E, and when obtained, reverts readily to Co^{2+}(aq) by oxidation of water (equation 71).

☐ Is there any observation in Experiments 1–3 which does *not* support the hypothesis?

■ Yes; hydrogen peroxide easily oxidises Fe^{2+}(aq) to Fe^{3+}(aq), but fails to oxidise Mn^{2+}(aq) to Mn^{3+}(aq). Mn^{2+}(aq) seems to be harder to oxidise than Fe^{2+}(aq).

To sum up, ignoring distinctions between thermodynamic and kinetic stability, the difficulty of the oxidation, or the stability of M^{2+}(aq) with respect to M^{3+}(aq), seems to vary in the order:

$$Cr^{2+}(aq) < Fe^{2+}(aq) < Mn^{2+}(aq) < Co^{2+}(aq) < Ni^{2+}(aq)$$

or, with elements in their order in the transition series:

$$Cr^{2+}(aq) < Mn^{2+}(aq) > Fe^{2+}(aq) < Co^{2+}(aq) < Ni^{2+}(aq)$$

There is therefore at least one point, between manganese and iron, where our hypothesis that it gets harder to oxidise M^{2+}(aq) in the sequence scandium to zinc seems in error. Nevertheless, in a general sense the hypothesis looks promising. Now let's test it quantitatively by looking at thermodynamic data.

9.2 Thermodynamic stability of dipositive and tripositive aqueous ions

SLC 14

A way of comparing thermodynamic stabilities of oxidised and reduced states of elements was introduced in a Second Level Course. For example, the thermodynamic stability of sodium and silver with respect to oxidation in water can be compared by considering the values of $\Delta G_m^\ominus$ for the following reactions:

$$Na(s) + H^+(aq) = Na^+(aq) + \tfrac{1}{2}H_2(g); \quad \Delta G_m^\ominus = -262 \text{ kJ mol}^{-1} \quad 72$$

$$Ag(s) + H^+(aq) = Ag^+(aq) + \tfrac{1}{2}H_2(g); \quad \Delta G_m^\ominus = 77 \text{ kJ mol}^{-1} \quad 73$$

Thus, in comparing the relative stabilities of M^{2+}(aq) and M^{3+}(aq), we could compare the values of $\Delta G_m^\ominus$ for the reaction

$$M^{2+}(aq) + H^+(aq) = M^{3+}(aq) + \tfrac{1}{2}H_2(g) \quad 74$$

TLC 1

for the elements of the first transition series. These values are recorded in the first of the two columns of data in Table 19. It is, however, more usual to carry out this comparison by using the standard *electrode potential* or standard redox potential, $E^\ominus$, so we shall examine both $\Delta G_m^\ominus$ and $E^\ominus$ values side by side. You will be familiar with redox potentials if you have studied another Third Level Chemistry Course. However, if you are unfamiliar with redox potentials, Appendix 1 provides the necessary introduction. What now follows is an application of that Appendix to redox systems containing ions of the type M^{2+}(aq) and M^{3+}(aq).

9.2.1 The redox potential $E^\ominus(M^{3+} | M^{2+})$

Values of $E^\ominus$ are quoted in volts, but they are easily obtained from the values of $\Delta G_m^\ominus$ for the corresponding redox reaction. In equation 74, the metal ion in the oxidised state is M^{3+}(aq), which appears on the right. However, $E^\ominus$ values refer to equations of this kind in which the oxidised state is *always* written on the left, so we first reverse equation 74:

$$M^{3+}(aq) + \tfrac{1}{2}H_2(g) = M^{2+}(aq) + H^+(aq) \quad 75$$

Because hydrogen gas and hydrogen ions always appear in such equations, tedious repetition is avoided by abbreviating equation 75 to

$$M^{3+}(aq) + e = M^{2+}(aq) \qquad 76$$

Here, the 'e' on the left-hand side is merely shorthand for '$[\frac{1}{2}H_2(g) - H^+(aq)]$'. It is not strictly an electron, when the symbol would have a superscript minus, e^-. However, by thinking of it as an electron, you can mark the fact that the total charge on the combination $[\frac{1}{2}H_2(g) - H^+(aq)]$ is -1, and confirm that equation 76 is balanced like equation 75.

The system in equation 76 is called a **couple**, and its **standard redox potential** is written $E^\ominus(M^{3+}|M^{2+})$, with the oxidised state to the left inside the bracket, and the reduced state to the right. This matches their positions in equation 76.

Now when $\Delta G_m^\ominus$ and $E^\ominus$ refer to the same reaction or couple with the oxidised state written on the left, the relationship between them is

$$\Delta G_m^\ominus = -nFE^\ominus \qquad 77$$

Here n is the coefficient of e on the left-hand side of the couple. It is also the decrease in oxidation state (always positive) between the oxidised state on the left of the couple and the reduced state on the right. F is a positive constant called the Faraday. The magnitude of F is such that to obtain $E^\ominus$ from $\Delta G_m^\ominus$ via equation 77, one divides $-\Delta G_m^\ominus/\text{kJ mol}^{-1}$ by $96.485n$. This gives $E^\ominus/\text{V}$, V being the symbol for the volt.

In the case of equations 75 and 76, $n = 1$, so if we write the respective $\Delta G_m^\ominus$ values as $\Delta G_m^\ominus(75)$ and $\Delta G_m^\ominus(76)$, then

$$\Delta G_m^\ominus(75) = \Delta G_m^\ominus(76) = -FE^\ominus(M^{3+}|M^{2+}) \qquad 78$$

Now equation 74 is the reverse of equation 75, so

$$\Delta G_m^\ominus(74) = -\Delta G_m^\ominus(75) \qquad 79$$

$$\Delta G_m^\ominus(74) = FE^\ominus(M^{3+}|M^{2+}) \qquad 80$$

$E^\ominus(M^{3+}|M^{2+})$ can therefore be calculated from $\Delta G_m^\ominus(74)$; both quantities appear alongside one another in Table 19. However, the value of $E^\ominus(Mn^{3+}|Mn^{2+})$ is missing.

SAQ 38 Calculate $E^\ominus(Mn^{3+}|Mn^{2+})$ from data in Table 19 and the earlier part of this Section. Pencil your answer into the Table.

Equation 80 shows that $\Delta G_m^\ominus(74)$ is proportional to $E^\ominus(M^{3+}|M^{2+})$, and you can see from Table 19 that both quantities are relatively large and positive when oxidation of $M^{2+}(aq)$ is difficult, or reduction of $M^{3+}(aq)$ is easy. For example, the experiments summarised in Figure 34 show that it is hard to oxidise $Co^{2+}(aq)$ to $Co^{3+}(aq)$.

☐ How is this reflected by the values of $\Delta G_m^\ominus(74)$ and $E^\ominus(M^{3+}|M^{2+})$ in Table 19?

■ For cobalt, both are large and positive. For the oxidation reaction akin to equation 74:

$$Co^{2+}(aq) + H^+(aq) = Co^{3+}(aq) + \tfrac{1}{2}H_2(g), \quad \Delta G_m^\ominus = 187 \text{ kJ mol}^{-1} \qquad 81$$

For the couple that refers to the reduction

$$Co^{3+}(aq) + e = Co^{2+}(aq), \quad E^\ominus = 1.94 \text{ V} \qquad 82$$

A list of $E^\ominus$ values in acid solution is given in the S343 *Data Book*. Roughly speaking, redox systems with $E^\ominus$ values greater (more positive) than about 1.1 V in acid solution contain powerful oxidising agents: conversely, those with $E^\ominus$ values less (more negative) than about -0.1 V in acid solution contain powerful reducing agents. *The more powerful the oxidising agent in a system, the more positive is $E^\ominus$.*

Now look again at the values of $E^\ominus(M^{3+}|M^{2+})$ in Table 19.

Table 19 Values* of $\Delta G_m^\ominus$ for equation 74, and of $E^\ominus(M^{3+}|M^{2+})$ for the first-row transition elements at 25 °C

Element	$\dfrac{\Delta G_m^\ominus(74)}{\text{kJ mol}^{-1}}$	$\dfrac{E^\ominus(M^{3+}|M^{2+})}{\text{V}}$
Sc	(−251)	(−2.6)
Ti	(−106)	(−1.1)
V	−24	−0.25
Cr	−41	−0.42
Mn	154	1.60
Fe	74	0.77
Co	187	1.94
Ni	405	(4.2)
Cu	444	(4.6)
Zn	675	(7.0)

* The values in parentheses are theoretical estimates.

☐ Does the thermodynamic stability of M^{2+}(aq) with respect to M^{3+}(aq) increase across the series? List any deviations.

■ The more positive $E^\ominus(M^{3+}|M^{2+})$, the more powerful an oxidising agent is the M^{3+}(aq) ion, and the harder it is to oxidise M^{2+}(aq). The negative values at the beginning of the series, and the positive values at the end show that overall it gets harder to oxidise M^{2+}(aq) from scandium to zinc. In fact, the thermodynamic stability of M^{2+}(aq) increases from element to element except from vanadium to chromium, where $E^\ominus$ falls very slightly from -0.25 V to -0.42 V, and between manganese and iron, where there is a much bigger fall from 1.60 V to 0.77 V. We detected this second exception, together with the overall increase in the stability of M^{2+}(aq), in the experiments of Figure 34.

Redox potentials can be used to predict whether or not a particular redox reaction is thermodynamically possible under standard conditions.*

> If we have two redox systems, and the $E^\ominus$ value for the first is more positive than that for the second, then the oxidised state in the first is thermodynamically capable of oxidising the reduced state in the second.

Make sure you understand this by trying SAQ 39.

SAQ 39 Use the information in Table 19 and the value $E^\ominus(S_2O_8^{2-}|2SO_4^{2-}) = 1.94$ V to find out which of the following reactions are thermodynamically favourable:

(a) oxidation of Cr^{2+}(aq) by Mn^{3+}(aq);
(b) oxidation of Mn^{2+}(aq) by Fe^{3+}(aq);
(c) oxidation of Mn^{2+}(aq) by Co^{3+}(aq);
(d) reduction of Co^{3+}(aq) by Fe^{2+}(aq);
(e) oxidation of Mn^{2+}(aq) by persulphate;
(f) reduction of persulphate by V^{2+}(aq).

9.2.2 Solvent decomposition

Apart from the actual values of $E^\ominus(M^{3+}|M^{2+})$, two other redox potentials are particularly relevant to the relative stabilities of the ions M^{2+}(aq) and M^{3+}(aq). They define, in a thermodynamic sense, how hard it is to either oxidise or reduce the solvent system, which, in acid solution, consists of water molecules and aqueous hydrogen ions. They are:

$$O_2(g) + 4H^+(aq) + 4e = 2H_2O(l); \quad E^\ominus = 1.23 \text{ V} \qquad 83$$

$$H^+(aq) + e = \tfrac{1}{2}H_2(g); \quad E^\ominus = 0.00 \text{ V} \qquad 84$$

The first defines the resistance of water to oxidation in acid solution, and it provides a reason why the ions Ni^{3+}(aq), Cu^{3+}(aq) and Zn^{3+}(aq) are unknown. The estimated values of $E^\ominus(M^{3+}|M^{2+})$ in Table 19 for these three metals are much greater than 1.23 V, so the aqueous M^{3+} ions are thermodynamically capable of oxidising water to oxygen. This probably provides a decomposition route of the type

$$4Cu^{3+}(aq) + 2H_2O(l) = 4Cu^{2+}(aq) + 4H^+(aq) + O_2(g) \qquad 85$$

which also happens to be fast, so the Cu^{3+}(aq) ions are never observed.

Now let's turn to the second redox potential, equation 84. It is zero, and defines the resistance to reduction of the hydrogen ions in the acid solutions that we have been using.

☐ Can you suggest why the ions Sc^{2+}(aq) and Ti^{2+}(aq) are unknown?

---

* You need not worry here about the proviso of standard conditions. This point will be taken up in Block 3.

■ The estimated values of $E^{\ominus}(M^{3+}|M^{2+})$ (and $\Delta G_m^{\ominus}(74)$) in Table 19 for both elements are negative, so $Sc^{2+}(aq)$ and $Ti^{2+}(aq)$ ions are thermodynamically capable of reducing aqueous hydrogen ions. This provides a decomposition route, which is presumably fast, so they are not observed:

$$Sc^{2+}(aq) + H^{+}(aq) = Sc^{3+}(aq) + \tfrac{1}{2}H_2(g) \qquad 86$$

Now let's inject our familiar note of caution about thermodynamic data such as electrode potentials. Compare the values of $E^{\ominus}(M^{3+}|M^{2+})$ for vanadium, chromium, manganese and cobalt with values of $E^{\ominus}$ for equations 83 and 84.

□ Do you detect an important limitation of predictions made using $E^{\ominus}$ values?

■ $V^{2+}(aq)$ and $Cr^{2+}(aq)$ are thermodynamically capable of reducing hydrogen ions, just as $Sc^{2+}(aq)$ and $Ti^{2+}(aq)$ do; $Mn^{3+}(aq)$ and $Co^{3+}(aq)$ are thermodynamically capable of oxidising water, just as $Cu^{3+}(aq)$ does. Yet $V^{2+}(aq)$, $Cr^{2+}(aq)$, $Mn^{3+}(aq)$ and $Co^{3+}(aq)$ can all be prepared in aqueous media. This is because these decompositions are usually very slow, although the decomposition rate of $Co^{3+}(aq)$ is significant. *Like all thermodynamic data, electrode potentials are subject to limitations imposed by kinetic considerations.*

SAQ 40 Titanium forms a black dichloride, $TiCl_2$. Predict what you would see when this compound is added to dilute acid in the absence of oxygen and write an equation for the reaction.

SAQ 41 In 1971 a black hydroxide of nickel in the oxidation state +3 was obtained with the formula NiO(OH). Predict what you would see when this compound is added to dilute sulphuric acid, and write an equation for the process.

SAQ 42 Besides defining the resistance of water to oxidation, the couple in equation 83 also defines the strength of oxygen (in contact with aqueous hydrogen ions) as an oxidising agent. How many of the ions $Cr^{2+}(aq)$, $Mn^{2+}(aq)$, $Fe^{2+}(aq)$ and $Co^{2+}(aq)$ are thermodynamically unstable to oxidation by oxygen in acid solution? How well does your conclusion square with the actual behaviour of these ions (as shown in Figure 34)?

9.3 Thermodynamic stability and electronic configuration

Let us now try to relate the increase in the stability of $M^{2+}(aq)$ with respect to $M^{3+}(aq)$ across the series to the electronic configuration of the ions. When $M^{2+}(aq)$ is converted to $M^{3+}(aq)$, it loses an electron. We have already seen in Section 8.2.2 that the radii of the M^{2+} ions decrease across the series. We related this to the dominant effect of the increase in nuclear charge over that of increasing electron repulsion as the electronic configuration of the M^{2+} ions changes from $[Ar]3d^0$ at calcium to $[Ar]3d^{10}$ at zinc.

□ Can you relate this dominance of increasing nuclear charge to the increasing stability of $M^{2+}(aq)$ with respect to oxidation, as discussed in Section 9.2.1?

■ The dominance of increasing nuclear charge suggests that the M^{2+} ions should lose a 3d electron more easily at the beginning of the series than at the end: this is consistent with our observation that at the beginning of the series $M^{2+}(aq)$ ions are so easily oxidised that they do not exist, and that at the end of the series the $M^{2+}(aq)$ ions are oxidised with such difficulty that $M^{3+}(aq)$ ions are not formed.

We can, however, go further than this by comparing the variation in $\Delta G_m^{\ominus}$ for the reaction

$$M^{2+}(aq) + H^{+}(aq) = M^{3+}(aq) + \tfrac{1}{2}H_2(g) \qquad 74$$

with the variation in the third ionisation energy of the metal, which involves loss of an electron from a gaseous M^{2+} ion:

$$M^{2+}(g) = M^{3+}(g) + e^{-}(g) \qquad 87$$

In both equation 74 and equation 87, an electron is lost from an M^{2+} ion.

Moving step by step across the transition series, it is just the identity of M that changes in both reactions: in both cases, the *variation* in energy is a *variation* in the amount of energy needed to remove an electron from the different M^{2+} ions. So although for any metal the values of $\Delta G_m^\ominus$ for equation 74 and the third ionisation energy will be different, we might expect that the two variations will *parallel* one another across the series. The values for $\Delta G_m^\ominus(74)$ for the first-row transition elements were recorded in Table 19 and they are shown as the green plot in Figure 35. The black plot shows the corresponding values of the ionisation energy of $M^{2+}(g)$, the third ionisation energy of the metals, I_3.

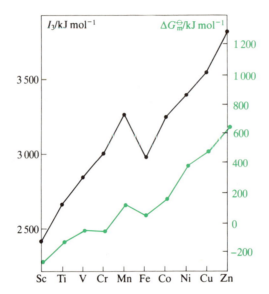

Figure 35 Third ionisation energies (black plot, left-hand axis) and $\Delta G_m^\ominus$ values (green plot, right-hand axis) for reaction 74, where M is a first-row transition element.

☐ Is our expectation of a parallelism fulfilled?

■ Yes: both plots show an overall increase*, thus showing that whether we deal with gaseous or aqueous M^{2+} ions, oxidation gets more difficult as we move across the series.

As described earlier in this Section, this increase can be related to the increasing nuclear charge. In both plots the overall increase is broken by the downward break from manganese to iron; this break will be explained in Block 3.

However, our expectations are not entirely fulfilled because there are some differences between the two variations: in particular the green plot looks like a diluted form of I_3, in which the overall increase and the half-way break have been very much toned down.

These differences are due to the fact that I_3 is associated with the loss of an electron from a gaseous M^{2+} ion, but reaction 74 involves the loss of an electron from an M^{2+} ion surrounded, as we saw in Section 5.5.5, by six water ligands. The differences between the two variations can be explained by considering the effects of the ligands on the energy levels of the M^{2+} ions. This exercise involves ligand-field theory, which is introduced in Block 2. Consequently, we defer discussion of this point until Block 3. For the moment, it is enough to note the important parallelism between the variations in $\Delta G_m^\ominus$ for equation 74, and those in I_3.

9.4 Trihalides

If the dihalides of a particular transition element can be prepared by the action of hydrogen halides on the metal, then the reaction between the metal and the appropriate halogen often yields a trichloride, tribromide or tri-iodide, if such compounds are capable of existence; for example

$$Fe(s) + \tfrac{3}{2}Cl_2(g) \xrightarrow{300-350\,°C} FeCl_3(s) \qquad 88$$

$$Cr(s) + \tfrac{3}{2}I_2(s) \xrightarrow{200-230\,°C} CrI_3(s) \qquad 89$$

* An *increase* in $\Delta G_m^\ominus$ is taken to mean that $\Delta G_m^\ominus$, whether it is initially positive or negative, becomes more positive; that is, the reaction is becoming less favourable. Conversely, a *decrease* in $\Delta G_m^\ominus$ is taken to mean that $\Delta G_m^\ominus$ becomes more negative.

The syntheses of MnF_3 and CoF_3 were described in Sections 4.3 and 5.4. Moreover, in addition to these dry methods, when a hydrated trihalide can be crystallised from a solution of the tripositive aqueous ion, dehydration by thionyl chloride is effective:

$$FeCl_3 \cdot 6H_2O(s) + 6SOCl_2(l) = FeCl_3(s) + 6SO_2(g) + 12HCl(g) \qquad 90$$

The pattern of known trihalides is shown in Table 20.

Table 20 The range of known trihalides of the first-row transition elements

Sc	Ti	V	Cr	Mn	Fe	Co	Ni	Cu	Zn
all known	all known	all known	all known	MnF_3 $MnCl_3$*	FeF_3 $FeCl_3$ $FeBr_3$	CoF_3	none known	none known	none known

* Decomposes above $-40\,°C$.

☐ What similarities do you detect between this pattern and that for the tripositive aqueous ions?

■ As with tripositive ions, trihalides are known only for the first seven elements.

In the particular case of the trichlorides, the unknown compounds are $CoCl_3$, $NiCl_3$, $CuCl_3$ and $ZnCl_3$. Since the dichlorides of these elements are well known, the four trichlorides are probably unstable to the decomposition reaction

$$MCl_3(s) = MCl_2(s) + \tfrac{1}{2}Cl_2(g) \qquad 91$$

Now consider the reverse of this reaction:

$$MCl_2(s) + \tfrac{1}{2}Cl_2(g) = MCl_3(s) \qquad 92$$

☐ What *overall* trend in $\Delta G_m^\ominus$ for this reaction is compatible with the information on trichlorides in Table 20?

■ The values of $\Delta G_m^\ominus$, and therefore the thermodynamic stability of the dichloride with respect to chlorination, should become increasingly positive across the series.

☐ Do you see any connection between this overall trend, and the overall trend in the thermodynamic stability of $M^{2+}(aq)$ with respect to $M^{3+}(aq)$?

■ In both equation 92 and

$$M^{2+}(aq) + H^+(aq) = M^{3+}(aq) + \tfrac{1}{2}H_2(g) \qquad 74$$

the dipositive oxidation state is oxidised to the tripositive, and there is an overall increase in $\Delta G_m^\ominus$ across the series.

☐ For reaction 74, this overall increase is broken by a decrease from manganese to iron. Is there any evidence in Table 20 of such a decrease in $\Delta G_m^\ominus$ for reaction 92?

■ Yes. $CrCl_3(s)$ is known at room temperature but $MnCl_3$ is not. This suggests that $\Delta G_m^\ominus$ increases from chromium to manganese as we expect: $MnCl_3$ is unknown at room temperature because it is unstable with respect to the dichloride and chlorine. However, $FeCl_3$ is both known and thermodynamically stable at room temperature. The stability of the trichloride thus increases from manganese to iron, which means that the stability of the dichloride, and $\Delta G_m^\ominus$ for reaction 92 decreases.

To show just how similar are the variations of $\Delta G_m^\ominus$ for reactions 74 and 92, the two are plotted together in Figure 36. The parallelism between the two variations is even closer than with the I_3 variation.

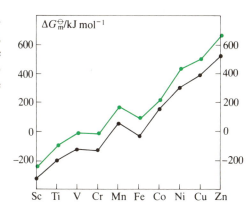

Figure 36 Values of $\Delta G_m^\ominus$ for reaction 74 (green), and for reaction 92 (black).

We can link the increasing stability of the dichlorides, in the same way as the stability of the M^{2+}(aq) ions, with the gradually tighter binding of electrons in the series of M^{2+} ions with configurations $[Ar]3d^n$. In making this link, we think in terms of dichlorides and trichlorides containing M^{2+} and M^{3+} ions, respectively.

Finally, Table 20 contains a reminder of a point made in Sections 4 and 5: among the trihalides of manganese and cobalt, only the trifluorides are stable with respect to reactions such as

$$MnX_3 = MnX_2 + \tfrac{1}{2}X_2 \qquad 93$$

$$CoX_3 = CoX_2 + \tfrac{1}{2}X_2 \qquad 94$$

at room temperature. Fluorine is most effective in stabilising the higher halide with respect to the lower. Iodine is least effective, as can be seen from the absence of FeI_3 in the column of iron trihalides. In fact the capacity of the halogens to bring out higher oxidation states in halogen compounds decreases in the order

SLC 15 $F > Cl > Br > I$. This order was explained in a Second Level Course.

9.5 Summary of Sections 8 and 9

1 In the first transition series, all known difluorides have the rutile structure, and all other dihalides (except those of zinc) have $CdCl_2$ or CdI_2 layer structures.

2 The ionic radii of M^{2+} ions decrease across the series but the decrease is not smooth. It follows a double-bowl shape with minima at V^{2+} and Ni^{2+}, and a cusp at Mn^{2+}.

3 M^{2+}(aq) ions are known for the last eight elements of the first transition series; M^{3+}(aq) ions are known for the first seven.

4 There is an overall increase in the stability of M^{2+}(aq) with respect to oxidation across the series, although an important decrease occurs from manganese to iron.

5 Unknown M^{2+}(aq) ions probably reduce water or H^+(aq) quickly; unknown M^{3+}(aq) ions probably oxidise water.

6 The stability variation of point 4 corresponds to the variation in thermodynamic stability: the values of $E^\ominus(M^{3+}|M^{2+})$ increase from scandium to manganese, fall from manganese to iron, and increase from iron to zinc.

7 This variation in $E^\ominus(M^{3+}|M^{2+})$ is equivalent to the variation in $\Delta G_m^\ominus$ for the reaction

$$M^{2+}(aq) + H^+(aq) = M^{3+}(aq) + \tfrac{1}{2}H_2(g) \qquad 74$$

The variation in $\Delta G_m^\ominus$ is a diluted version of the variation in the third ionisation energy of the elements.

8 Some of the known M^{2+}(aq) and M^{3+}(aq) ions are thermodynamically unstable in aqueous solution; they exist because their reactions with the solvent are very slow.

9 All dihalides of the first-row transition elements are known except the dihalides of scandium, TiF_2 and CuI_2.

10 At room temperature, trifluorides of the first seven elements in the first transition series are known, trichlorides and tribromides of the first six (except for $MnCl_3$ and $MnBr_3$) and tri-iodides of the first four.

11   The values of $\Delta G_m^\ominus$ for halide reactions of the type

$$MX_2(s) + \tfrac{1}{2}X_2 = MX_3(s) \qquad\qquad 95$$

increase from scandium to manganese, decrease to iron and then increase to zinc. This variation is very similar to that in $\Delta G_m^\ominus$ for reaction 74, and is again a diluted version of the variation in the third ionisation energy of the elements.

12   The capacity of the halogens to bring out higher oxidation states in halogen compounds decreases in the order F > Cl > Br > I.

SAQ 43   What evidence is there in Table 20 to suggest that chlorine is better than bromine or iodine at bringing out high oxidation states?

10 FINAL COMMENTS ON OXIDATION STATES +2 AND +3

We have now considered five systems in which the stability of the dipositive oxidation state with respect to the tripositive oxidation state follows a similar pattern across the first transition series. The five systems are the aqueous ions, the fluorides, the chlorides, the bromides and the iodides, and in each case, the stability of the dipositive oxidation state shows a general increase across the series, although there is a marked decrease from manganese to iron.

Such a similarity shows that, in the case of the simple compounds examined in this Block, when we classify compounds by oxidation state we can make useful generalisations about their chemistry. To some extent, these generalisations can even be of an absolute kind. Thus, nowhere in the systems that we have examined has a compound of copper(III) or zinc(III) appeared, and we might therefore feel justified in saying that compounds of this type have low stability with respect to copper(II) or zinc(II). Such a conclusion would be correct: no compound of zinc(III) has ever been made, and it is very difficult to make compounds of copper(III) by oxidising copper(II). However, you must always remember that generalisations of this kind are statements of probability only, and that the stability of a particular oxidation state can be affected by manipulating the chemical environment in which it occurs. You have met examples of this in this Block. Thus, at room temperature, MnF_3 is stable with respect to MnF_2, but the unknown compound MnI_3 is almost certainly unstable with respect to MnI_2 and iodine.

This is just one example of the fact that though generalisations about the stabilities of oxidation states are often useful as a first approximation, the synthesis of new or unusual oxidation states requires an ability to defeat the kind of pessimistic generalisation that we made before about zinc(III) and copper(III). Some of the synthetic methods that can be used are described in Section 13.

SAQ 44   Six of the ten first-row transition elements form well-characterised oxides, M_2O_3. Which six do you think they are? A seventh is of borderline stability and loses oxygen at a considerably lower temperature than the others when it is heated in air. Which one do you think it is? Which halogen does oxygen most resemble in its ability to bring out high oxidation states?

11 LOWER OXIDATION STATES

In Sections 6 to 10 we have restricted ourselves to the study of aqueous ions, halides, oxides and salts formed with common acids. With this restriction the only important oxidation state below +2 for the first transition series is +1. At normal temperatures, copper is the only first-row transition metal to form compounds in this oxidation state. We shall use the chemistry of copper to introduce the properties of *potential diagrams*. These are a valuable way of representing the thermodynamics of the redox reactions of an element in aqueous solution. Once introduced in this way, we shall then use them for other elements.

11.1 Copper

The **potential diagram** for copper in acid solution is given below:

$$Cu^{2+} \xrightarrow{0.16} Cu^{+} \longrightarrow Cu$$
$$\underline{\phantom{Cu^{2+}}\quad 0.34 \quad\phantom{Cu^{+}}}\uparrow$$

The most prominent species in the various oxidation states are written down in a chain with the most oxidised state on the left and the most reduced on the right. The two oxidation states in any couple are connected by an arrow pointing from left to right, and this is labelled with the standard electrode potential of that couple. Strictly speaking, the values should be followed by volt symbols, V, to indicate their units, but we omit these to make the diagrams less cluttered. The potential diagram here is not complete because it contains only the values of $E^{\ominus}(Cu^{2+}|Cu)$ and $E^{\ominus}(Cu^{2+}|Cu^{+})$: $E^{\ominus}(Cu^{+}|Cu)$ is missing. However, we can complete it by using a rule that we state here without proof. The proof and further details about potential diagrams are given in Appendix 2. The rule is:

> Between two species in any cycle on a potential diagram, the sum of the values of $nE^{\ominus}$ in a clockwise direction is equal to the sum of the values of $nE^{\ominus}$ in an anticlockwise direction. n is the decrease in oxidation state (always positive) between the oxidised and the reduced species in a couple.

Now let's apply this rule to the potential diagram for copper. Suppose we move from Cu^{2+} to Cu.

☐ What is the sum of $nE^{\ominus}$ in a clockwise direction?

■ It is $(0.16\ V + E^{\ominus}(Cu^{+}|Cu))$, n being one for both steps.

☐ What is the sum of $nE^{\ominus}$ in an anticlockwise direction?

■ It is $2 \times 0.34\ V$ or $0.68\ V$.

☐ What is the value of $E^{\ominus}(Cu^{+}|Cu)$?

■ $0.16\ V + E^{\ominus}(Cu^{+}|Cu) = 0.68\ V$

$E^{\ominus}(Cu^{+}|Cu) = 0.52\ V$

You can now insert this value on the potential diagram. We now know the redox potentials of the couples connecting an intermediate state Cu^{+}, with an upper state Cu^{2+} and a lower state Cu.

Notice that the $E^{\ominus}$ value involving Cu^{+} and the lower state, $E^{\ominus}(Cu^{+}|Cu)$, is greater (by 0.36 V) than the $E^{\ominus}$ value involving the upper state and Cu^{+}, $E^{\ominus}(Cu^{2+}|Cu^{+})$.

☐ What redox reaction is therefore thermodynamically favourable?

■ Cu^+ going to Cu will oxidise Cu^+ to Cu^{2+}. In other words, the potential diagram tells us that $Cu^+(aq)$ is thermodynamically unstable with respect to the reaction

$$2Cu^+(aq) = Cu(s) + Cu^{2+}(aq) \qquad 96$$

☐ What kind of reaction is this?

■ It is a disproportionation. $Cu^+(aq)$ is thermodynamically unstable with respect to disproportionation.

The result is quite general:

> Whenever the couple linking an ion or compound with a lower oxidation state has an $E^\ominus$ value greater than that of the couple linking it with an upper oxidation state, the ion or compound is thermodynamically unstable to disproportionation.

In this case, $Cu^+(aq)$ does indeed decompose according to equation 96. However, copper(I) compounds can be stabilised if the equilibrium is shifted to the left by adding an anion that forms an insoluble compound with $Cu^+(aq)$.

For example, if aqueous iodide ions are added to a blue solution of $Cu^{2+}(aq)$, a white precipitate is formed and a deep brown colour appears in the solution. This is iodine, which means that the iodide has been oxidised by something. The white precipitate is CuI, the iodide of copper(I), formed by reduction of $Cu^{2+}(aq)$:

$$Cu^{2+}(aq) + 2I^-(aq) = CuI(s) + \tfrac{1}{2}I_2(s) \qquad 97$$

If excess iodide is present, the iodine dissolves to form the complex ion, $I_3^-(aq)$.

The presence of $I^-(aq)$ precipitates CuI, which is sufficiently insoluble to be stable with respect to disproportionation in aqueous media.

As well as CuI, the insoluble salts CuBr and CuCl can also be made from aqueous solution. They are conveniently prepared by reducing a solution containing a mixture of copper(II) sulphate and sodium bromide or chloride with a stream of SO_2; for example

$$2Cu^{2+}(aq) + 2Br^-(aq) + SO_2(g) + 2H_2O(l) = 2CuBr(s) + SO_4^{2-}(aq) + 4H^+(aq) \qquad 98$$

CuF cannot be prepared in this way. Indeed, it is unknown at room temperature, but can be made by heating CuF_2, which first melts and then decomposes at about 1 000 °C. On cooling, however, CuF decomposes by disproportionation:

$$2CuF(s) = CuF_2(s) + Cu(s) \qquad 99$$

Apart from CuCl, CuBr and CuI, no first-row transition-metal monohalides are known at room temperature. The behaviour of CuF gives a clue to the type of decomposition reaction to which they are unstable.

Many simple compounds of copper(I) show a marked tendency to revert to copper(II). The red copper(I) oxide, Cu_2O, can be made by boiling an alkaline suspension of $Cu(OH)_2$ with the reducing agent hydrazine, N_2H_4, but on standing in the atmosphere, Cu_2O slowly forms the black copper(II) oxide, CuO.

Copper(I) sulphate, Cu_2SO_4, can be made by heating copper(I) oxide with dimethyl sulphate at 160 °C:

$$Cu_2O(s) + (CH_3)_2SO_4(l) = Cu_2SO_4(s) + (CH_3)_2O(g) \qquad 100$$

This unusual method is necessary because it avoids the use of water: on contact with water, Cu_2SO_4 is not sufficiently insoluble to be stabilised with respect to disproportionation, and it instantly forms a solution of copper(II) sulphate and a precipitate of metallic copper.

11.2 Summary of Section 11

1 The thermodynamics of the redox chemistry of transition elements in aqueous solution can be concisely expressed by potential diagrams.

2 Among the first-row transition elements, copper is the only element to form simple compounds like halides, salts or oxides in the +1 oxidation state.

3 The behaviour of these compounds suggests that if similar compounds could be made for other first-row transition metals, they would quickly revert to higher oxidation states, either by aerial oxidation or by disproportionation.

SAQ 45 According to Section 8.2 and Table 16, CuI_2 is unknown. What reasonable decomposition reaction can you suggest?

SAQ 46 Copper does not react with dilute acids to give hydrogen. What properties of the potential diagram for copper are consistent with this fact?

SAQ 47 The compounds Ni_2O and VI are unknown, even in the absence of air. Suggest reasonable decomposition routes.

12 HIGHER OXIDATION STATES IN AQUEOUS SOLUTION

In Sections 6–11 we have so far concentrated on oxidation states of +3 or less, and in particular, on oxidation states +2 and +3. We now introduce some important examples of oxidation states greater than +3.

12.1 Chromium

The potential diagram for chromium in acid solution is given below:

$$Cr_2O_7^{2-} \xrightarrow{1.36} Cr^{3+} \xrightarrow{-0.42} Cr^{2+} \xrightarrow{-0.90} Cr$$

In aqueous solution, the important higher oxidation state of chromium is +6; in acid, it occurs as the orange dichromate(VI) ion, $Cr_2O_7^{2-}$, unless the solution is very dilute when the dominant species present is the ion $HCrO_4^-$(aq).

The dichromate(VI) ion is a strong oxidising agent:

$$Cr_2O_7^{2-}(aq) + 14H^+(aq) + 6e = 2Cr^{3+}(aq) + 7H_2O(l); \quad E^\ominus = 1.36 \text{ V} \quad \textbf{101}$$

For example, it oxidises Fe^{2+}(aq) to Fe^{3+}(aq):

$$Fe^{3+}(aq) + e = Fe^{2+}(aq); \quad E^\ominus = 0.77 \text{ V} \quad \textbf{102}$$

and acidified hydrogen peroxide to oxygen:

$$O_2(g) + 2H^+(aq) + 2e = H_2O_2(aq); \quad E^\ominus = 0.69 \text{ V} \quad \textbf{103}$$

During the reaction with hydrogen peroxide:

$$Cr_2O_7^{2-}(aq) + 8H^+(aq) + 3H_2O_2(aq) = 2Cr^{3+}(aq) + 3O_2(g) + 7H_2O(l) \quad \textbf{104}$$

the orange dichromate(VI) is converted first to a deep blue intermediate, but eventually to greenish-violet Cr^{3+}(aq).

Because dichromate(VI) is a strong oxidising agent, even stronger oxidising agents such as persulphate are needed to convert chromium(III) to chromium(VI) in acid solution.

However, the situation is different in alkaline solution. Look again at reaction 104 and think of it as an equilibrium system.

☐ If OH^-(aq) is added, in which direction does the equilibrium move?

- By removing $H^+(aq)$, hydroxide ions shift the equilibrium to the left, and stabilise chromium(VI).

□ Do you remember another effect that $OH^-(aq)$ has on the dichromate ion, $Cr_2O_7^{2-}(aq)$?

SFC 6 ■ In the Science Foundation Course, you encountered the conversion of orange dichromate(VI) to yellow chromate(VI) by $OH^-(aq)$:

$$Cr_2O_7^{2-}(aq) + 2OH^-(aq) = 2CrO_4^{2-}(aq) + H_2O(l) \qquad 105$$

This change is reversed by acid.

□ Is the interconversion of dichromate and chromate a redox reaction?

■ No; the oxidation state of chromium is +6 in both ions.

These observations suggest that in alkali, chromium(VI) may be stabilised in the form of $CrO_4^{2-}(aq)$. This is correct. When sodium hydroxide is added to $Cr^{3+}(aq)$, a green precipitate of chromium(III) hydroxide, $Cr(OH)_3$, is formed. If this is heated with aqueous hydrogen peroxide, a yellow solution of $CrO_4^{2-}(aq)$ is obtained:

$$2Cr(OH)_3(s) + 3H_2O_2(aq) + 4OH^-(aq) = 2CrO_4^{2-}(aq) + 8H_2O(l) \qquad 106$$

Here chromium(III) has been oxidised to chromium(VI). However, if the solution is acidified while excess H_2O_2 is still present, $CrO_4^{2-}(aq)$ is converted to $Cr_2O_7^{2-}(aq)$ and the latter then decomposes via equation 104: in acid solution, chromium(VI) is reduced to chromium(III). The whole episode is summarised in Figure 37 and shown in the second sequence of Videocassette 1. It provides another illustration of an observation made in Section 5.3: high oxidation states are often stabilised in alkaline media.

Chromium(VI) is the highest known oxidation state of chromium. It shares a characteristic with the highest oxidation states of titanium and manganese that was noted in Sections 3.5 and 4.4.3: it is equal to the number of outer electrons in the free atom ($3d^5 4s^1$).

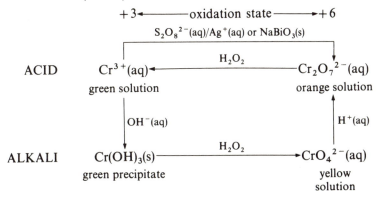

Figure 37 Interconversion reactions of chromium(III) and chromium(VI) in acid and alkaline solution. Note that H_2O_2 converts chromium(VI) to chromium(III) in acid, but converts chromium(III) to chromium(VI) in alkali.

12.2 Manganese

Our discussion of the oxidation states of manganese will revolve around the potential diagram for this element, so to begin with, do SAQ 48.

SAQ 48 A potential diagram for manganese in acid solution is shown below:

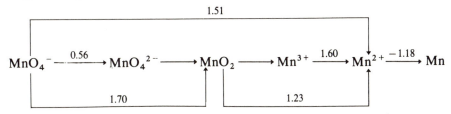

The diagram is incomplete. Complete it by calculating $E^\ominus(MnO_4^{2-}|MnO_2)$ and $E^\ominus(MnO_2|Mn^{3+})$.

The reactions of the oxidation states of manganese in aqueous solution were discussed in Sections 4.4 and 4.7, where we concentrated especially on the strong oxidising powers of permanganate(VII), MnO_4^-(aq). These strong oxidising powers are apparent in the large positive values of $E^\ominus(MnO_4^-|Mn^{2+})$ and $E^\ominus(MnO_4^-|MnO_2)$ in the potential diagram. You should now do the following SAQs to show that the potential diagram that you have just completed is consistent with other reactions of manganese in aqueous solution.

SAQ 49 The persulphate ion is a powerful oxidising agent:

$$S_2O_8^{2-}(aq) + 2e = 2SO_4^{2-}(aq); \quad E^\ominus = 1.94 \text{ V} \qquad 107$$

Are the values of $E^\ominus$, and those in the potential diagram consistent with the observation that in acid solution, persulphate will oxidise Mn^{2+}(aq) to Mn^{3+}(aq), and then Mn^{3+}(aq) to MnO_4^-(aq)?

SAQ 50 Unless the concentration of Mn^{3+}(aq) is small, this ion is unstable in acid and decomposes:

$$2Mn^{3+}(aq) + 2H_2O(l) = Mn^{2+}(aq) + MnO_2(s) + 4H^+(aq) \qquad 30$$

Another manganese species that is unstable in acid is the green manganate(VI) ion, MnO_4^{2-}(aq). It can be made in alkali, but in acid it decomposes to the brown MnO_2 and purple MnO_4^-(aq):

$$3MnO_4^{2-}(aq) + 4H^+(aq) = 2MnO_4^-(aq) + MnO_2(s) + 2H_2O(l) \qquad 36$$

Show that the occurrence of these two decomposition reactions is consistent with the potential diagram for manganese in acid solution.

12.3 Iron

The only oxidation state of iron greater than $+3$ that is discussed in this Block is introduced in SAQs 51 and 52.

SAQ 51 When chlorine is passed through a suspension of iron(III) hydroxide in KOH solution, a red solution containing the ferrate(VI) ion, FeO_4^{2-}(aq), is obtained. From the solution, the compound K_2FeO_4 can be crystallised, but when this is dropped into dilute acid, a yellow–brown solution of Fe^{3+}(aq) is formed and oxygen is evolved. Are these observations consistent with the idea that high oxidation states are stabilised in alkali? Write an equation for the reaction between K_2FeO_4 and dilute acid.

SAQ 52 A potential diagram for iron in acid solution is shown below

$$FeO_4^{2-} \xrightarrow{2.2} Fe^{3+} \xrightarrow{0.77} Fe^{2+} \xrightarrow{-0.46} Fe$$

To what extent, if at all, is the potential diagram consistent with the observations (a)–(f) below.

(a) When potassium ferrate(VI) is dissolved in dilute acid, the red FeO_4^{2-} ion is reduced to the yellow–brown Fe^{3+}(aq) ion, and oxygen is evolved.

(b) Atmospheric oxygen oxidises an acid solution of the green Fe^{2+}(aq) ion to the yellow–brown Fe^{3+}(aq).

(c) The oxidation cited in (b) is very slow.

(d) Iron dissolves in dilute HCl or H_2SO_4, forming Fe^{2+}(aq) and evolving hydrogen.

(e) When hydrogen gas is bubbled through a solution of Fe^{3+}(aq), reduction to Fe^{2+}(aq) does not occur.

(f) Fe^{2+}(aq) does not disproportionate to Fe^{3+}(aq) and metallic iron.

12.4 Structural chemistry of the higher oxidation states of chromium, manganese and iron

X-ray diffraction studies of the compounds K_2CrO_4, K_2MnO_4 and K_2FeO_4 show that all three compounds contain groupings in which a metal atom is surrounded by four oxygens at the corners of a regular tetrahedron. If these are thought of as MO_4^{2-} ions, then the geometry of the chromate(VI) ion is that shown in Figure 38a. It can be interpreted with a set of *resonance structures*, one of which is shown in Figure 38b; each oxygen has a complete octet.

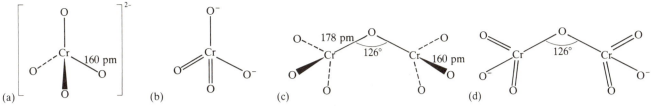

Figure 38 (a) Geometry of the chromate(VI) ion; (b) resonance structure for the chromate(VI) ion; (c) geometry of the dichromate(VI) ion; (d) resonance structure for the dichromate(VI) ion.

SLC 17 **SAQ 53** (*revision*) In a Second Level Course, you were able to predict the shapes of halides and oxoanions of the typical elements by applying valence-shell electron-pair repulsion theory to the outer s and p electrons. Assume that this theory can be similarly applied to the s and d electrons of the transition elements. What shapes would you predict for the ions CrO_4^{2-} and FeO_4^{2-}? Check your predictions against the results of the X-ray diffraction studies. How well does the theory cope with these examples from transition-metal chemistry?

In the chemistry of the typical elements, the shapes of molecules provide evidence of the repulsive effect of non-bonding electrons, and this is taken account of in a systematic way in valence-shell electron-pair repulsion theory. In transition-metal chemistry, the molecular shapes provide no sustained evidence of this type, and the theory is unreliable.

The tetrahedral coordination around chromium in the chromate(VI) ion is also evident in the dichromate(VI) ion. In $(NH_4)_2Cr_2O_7$, two CrO_4 tetrahedra link up as shown in Figure 38c. The central Cr—O bonds are longer than the terminal ones. This is consistent with the resonance structure of Figure 38d, which implies that the terminal bonds have the higher bond order.

13 SUMMARY OF OXIDATION-STATE PATTERNS IN THE FIRST TRANSITION SERIES

In Figure 39 we summarise those oxidation states of the elements from potassium to zinc which are found in stoichiometric binary halides and other simple salts, in anhydrous binary oxides, and in aqueous monatomic cations and aqueous oxoanions.

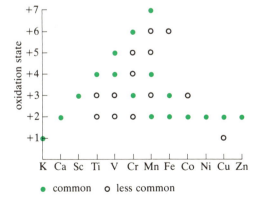

Figure 39 Oxidation-state pattern for the elements from potassium to zinc, compiled from a survey of aqueous monatomic ions and aqueous oxoanions, stoichiometric anhydrous oxides, halides and simple salts.

The ligands bound to the transition element in these substances are just F^-, Cl^-, Br^-, I^-, O^{2-} and H_2O, plus anions such as SO_4^{2-}, NO_3^- and ClO_4^-, which also bind through oxygen. Yet even with this narrow range of ligands, a wide range of oxidation states is achieved. Indeed, Figure 39 does not reveal the full range that is possible, because the substances to which it has been restricted exclude solid fluoride and oxide complexes. In Section 5.7, you saw that it was possible to make two compounds of this type, Cs_2CoF_6 and Cs_3CoO_4, which contain the high oxidation states cobalt(IV) and cobalt(V). The reactions used are, in fact, examples of a general method of synthesising high oxidation states—the action of hot fluorine and oxygen on solids containing lower oxidation states. For example, in the second sequence of Videocassette 1 you can see the preparation of the solid nickel(IV) and copper(III) complexes, K_2NiF_6 and K_3CuF_6, by these methods, along with demonstrations of their powerful oxidising properties. Likewise, the complex copper(III) oxide, $KCuO_2$, can be made by heating a mixture of K_2O and CuO in oxygen:

$$K_2O(s) + 2CuO(s) + \tfrac{1}{2}O_2(g) = 2KCuO_2(s) \qquad 108$$

Finally, by fluorinating a mixture of caesium and copper(II) chlorides in the right proportions with fluorine at a pressure of 350 atm, it is even possible to obtain the orange-red copper(IV) compound, Cs_2CuF_6. Thus, if we admit solid oxide and fluoride compounds into the range of substances covered by Figure 39, the upper threshold is raised: with this change, Figure 39 becomes Figure 40.

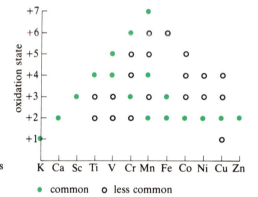

Figure 40 The pattern of Figure 39, modified by extending it to include oxidation states found in complex fluorides and oxides such as Cs_2CoF_6 and $KCuO_2$. This raises the upper threshold at the right-hand end of the series.

In both Figures, the green circles denote the most common oxidation states of each element. Unlike the compounds of oxidation states marked by open circles, these compounds are usually rather stable with respect to oxidation, reduction and disproportionation.

Oxidation states *higher* than those in Figure 40 have not been made, even with ligands other than the restricted set, listed at the beginning of this Section, to which the Figure applies. However, as you will see in Blocks 4 and 6, there is a new class of ligands which can break through the lower limits of Figure 40, and generate new *low* oxidation states. For example, one such ligand is phosphorus trifluoride, PF_3. If this is heated with CoI_2, copper and hydrogen under pressure, a yellow liquid hydride complex $[Co^IH(PF_3)_4]$ is produced. When dissolved in ether, the liquid is reduced by amalgamated potassium to colourless crystals of $K[Co(PF_3)_4]$, a compound of cobalt in oxidation state -1. You will encounter many similar instances in Blocks 4 and 6.

 We recommend that you now view the second sequence of Videocassette 1, referring to the S343 Audiovision Booklet for guidance. It shows a series of experiments on high oxidation states of the first-row transition elements.

13.1 The overall profile of oxidation states

Figure 40 shows that in the first half of the transition series from scandium to manganese, the range of oxidation states increases gradually, and the highest oxidation state is equal to the number of outer electrons. In the second half, this is not so, and the range of oxidation states becomes more restricted.

Let us first consider why the highest oxidation state never *exceeds* the number of outer electrons. Chromium atoms have six outer electrons, and in the chromate ion, CrO_4^{2-}, the element attains its highest oxidation state of $+6$. Suppose now that chromium tries to exceed this by forming CrO_4^-, the analogue of permanganate(VII), which, if it did exist, would contain chromium(VII). Why does it not succeed?

Do the following SAQ now.

SAQ 54 Draw one of the resonance structures for the CrO_4^- ion. How many electrons must chromium supply to form the bonds? Are the valence electrons sufficient? If not, why will it be difficult to find extra electrons? If CrO_4^- did exist, what kind of redox behaviour would it display?

So the creation of CrO_4^- would require the promotion of one of the inner core electrons to an outer 3d, 4s or 4p orbital, where it could be shared with the oxygen atoms. This requires so much energy that CrO_4^- cannot exist.

Now let us turn to the question of why, beyond manganese, it is not even possible to make compounds in which the oxidation state is *equal* to the number of outer electrons. For example, although $KMnO_4$ is known, $KFeO_4$, which contains iron(VII), is not.

☐ Would the bonding in this compound require the promotion of electrons from the argon core?

■ No; as with CrO_4^- and MnO_4^-, seven electrons must be contributed by the metal atom to form the bonds. Iron has eight valence electrons outside the argon core.

Likewise, iron does not form the oxide FeO_4, which would require all eight valence electrons of the metal atom. These examples show that in the second half of the series, beyond manganese, not even all the *valence* electrons are available to form bonds.

Now in earlier Sections in this Block, you have seen the way in which the increase in nuclear charge dominates over the effect of increase in interelectronic repulsion across the series. For example, it was apparent in the overall contraction in the radii of M^{2+} ions (Section 8.2.2) and in the increasing stability of $M^{2+}(aq)$ ions with respect to oxidation (Section 9.2.1). It is this increase in nuclear charge which binds the valence electrons more tightly in the second half of the series, and which makes it impossible to synthesise compounds such as FeO_4 in which all the valence electrons are used in bonding.

It is even possible to see that of the two types of valence electron in the free atoms, the problem lies with the 3d electrons rather than with 4s. At the beginning of the series the scandium atom has the configuration $[Ar]3d^14s^2$. Here, as at the preceding element calcium, the 4s level is filled and must have a lower energy than the 3d, since the electrons occupy the level of lowest energy first. Now let us move on towards the end of the series.

☐ From Table 1, the configuration of the copper atom is $[Ar]3d^{10}4s^1$. Which of the two levels, 3d and 4s, is now the lower?

■ Here 3d has been filled before 4s: 3d now has the lower energy of the two.

So as one moves across the series, the energy of the 3d level is lowered relative to that of the 4s. Or, putting it in a different way, the 3d electrons in the atom become more tightly bound by the increasing nuclear charge than do the 4s.

☐ How is this consistent with the observed oxidation states of zinc in Figure 40?

■ The maximum, and only oxidation state of zinc found in compounds is +2. Only the two 4s electrons on the zinc atom are available for bonding.

At zinc, the ten 3d electrons have become so stabilised that they are, for all chemical purposes, part of the inner core of electrons, which is not available for bonding. In many respects, therefore, zinc behaves like an alkaline earth metal.

13.2 Summary of Section 13

1 The maximum oxidation state follows the number of valence electrons from potassium to manganese.

2 Beyond manganese, the maximum oxidation state drops increasingly below the number of valence electrons, probably because the latter become more tightly bound by the increasing nuclear charge.

3 As the nuclear charge increases, 3d electrons become more tightly bound than 4s electrons. This reveals itself most notably in the chemistry of zinc, whose compounds are confined to the $+2$ oxidation state.

SAQ 55 The following solid compounds have not yet been prepared: K_2VO_4, K_2NiO_4, $ScCl_2$, Fe_2O_7, NiF_5, TiF_2. For each compound, which of the following statements do you think can best account for the failure to prepare it?

(a) In the first transition series, the electrons in the argon core are not available for bonding.

(b) As one moves across the first transition series, the valence electrons gradually become more tightly bound.

(c) The compound is unstable with respect to the metal and a compound in a higher oxidation state.

14 SUMMARY OF BLOCK 1

This Block has used the chemistry of some simple compounds and aqueous ions of the first-row transition elements to elicit generalisations that can be made about transition-element chemistry. Some of the most important generalisations that were discussed are listed as points 1–17 below.

First we had a series of statements that carry information about the first-row transition elements in a highly condensed form:

1 The last eight elements in the first transition series form $M^{2+}(aq)$ ions.

2 All dihalides of the first-row transition metals are known except the dihalides of scandium, TiF_2 and CuI_2.

3 In the first transition series, compounds of similar formula type often have a similar structure. Thus, all difluorides have a rutile type of structure, and all dichlorides, dibromides and di-iodides except those of zinc (four coordinate), have some form of CdX_2 layer structure.

4 The first seven elements of the first transition series form $M^{3+}(aq)$ ions.

5 In the case of aqueous ions, halides and oxides, there is a general tendency for the stability of the $+2$ oxidation state with respect to the $+3$ oxidation state to increase across the series.

6 The maximum attainable oxidation state, which is found in fluorine and/or oxygen compounds, follows the number of valence electrons from potassium to manganese, but beyond manganese it drops increasingly below the number of valence electrons as the electrons become more tightly bound.

7 Copper is the only element to exhibit the $+1$ oxidation state in an aqueous ion and in simple compounds like halides, oxides and salts. For other elements the $+1$ oxidation state in these forms is unstable with respect to disproportionation.

Then we had a series of comments that apply to the transition elements as a whole:

8   All transition elements are metals.

9   Strong bonds formed by d electrons are mainly responsible for the high melting temperatures, boiling temperatures and $\Delta H_f^\ominus(M, g)$ of the transition metals.

10   Those elements whose free atoms have the largest numbers of unpaired d electrons have, *for the most part*, the highest melting temperatures, boiling temperatures and cohesive energies.

11   The transition elements are more alike than the typical elements, notably in the formation of compounds of similar formula type by consecutive elements.

12   Transition-metal ions and compounds are often coloured, and in many cases this seems to be associated with the presence of unpaired d electrons.

13   The oxidation states of a transition element often differ by one.

14   The elements of a transition series behave more similarly in reactions in which the oxidation state is conserved than in oxidation–reduction reactions.

15   Valence-shell electron-pair repulsion theory cannot be applied to compounds of the transition elements as reliably as it can for compounds of the typical elements.

Finally, there were two generalisations that do, in fact, apply equally well to both the transition elements and other metals:

16   High oxidation states are often stabilised in alkaline media.

17   The capacity of the halogens to bring out high oxidation states varies in the order $F > Cl > Br > I$.

15 VANADIUM: A FINAL CASE STUDY

Vanadium is a very colourful element. It is commonly found in laboratories as the yellow compound sodium vanadate(V), $NaVO_3$. This dissolves in acid to give a solution of vanadium(V) as $VO_2^+(aq)$:

$$NaVO_3(s) + 2H^+(aq) = Na^+(aq) + VO_2^+(aq) + H_2O(l) \qquad 109$$

In the second sequence of Videocassette 1 you can see the reduction of $VO_2^+(aq)$ by zinc, which generates a sequence of lower oxidation states. Along with other observations on the chemistry of vanadium, this forms the substance of the last SAQ, SAQ 56, which tests your grasp of the generalisations about the first-row transition elements that we have made in this Block.

SAQ 56   The sequence of colours observed during reduction of $VO_2^+(aq)$ is yellow, green, blue, green and finally a steel-blue or lavender colour. The first green colour observed is a mixture of the starting material, $VO_2^+(aq)$ with the blue colour of the ion $VO^{2+}(aq)$. The latter is then reduced to the green $V^{3+}(aq)$, which, with some difficulty, can be reduced to the lavender $V^{2+}(aq)$. On standing after boiling, air oxidises $V^{2+}(aq)$ back to $V^{3+}(aq)$. Halides are also formed in all the oxidation states studied here. VF_5, a white solid, is made by fluorination of the metal. It melts at $20\,°C$, and is the only vanadium pentahalide known, but VF_4, a green solid, VCl_4, a red-brown liquid containing regular tetrahedral molecules made by chlorination of the metal, and VBr_4, a purple unstable liquid, have all been prepared. VI_4 is not known but as mentioned in Sections 8.2 and 9.4, all vanadium dihalides and trihalides are known. VF_2 has the rutile structure; VCl_2, VBr_2 and VI_2 have CdX_2 layer structures.

How many of the generalisations 1–17 in Section 14 are in evidence in this short description of vanadium chemistry?

APPENDIX 1 STANDARD REDOX POTENTIALS

This Section should be read in parallel with Section 9.2.1.

SLC 18 In a Second Level Course, we introduced a way of comparing the thermodynamic stabilities of metals with respect to oxidation into their aqueous ions. For example, in comparing sodium, magnesium, aluminium and silver, we wrote down the following reactions in order of increasing $\Delta G_m^\ominus$:

Reaction	$\Delta G_m^\ominus / \text{kJ mol}^{-1}$
$Na(s) + H^+(aq) = Na^+(aq) + \frac{1}{2}H_2(g)$	-261.9
$\frac{1}{2}Mg(s) + H^+(aq) = \frac{1}{2}Mg^{2+}(aq) + \frac{1}{2}H_2(g)$	-227.4
$\frac{1}{3}Al(s) + H^+(aq) = \frac{1}{3}Al^{3+}(aq) + \frac{1}{2}H_2(g)$	-161.7
$Ag(s) + H^+(aq) = Ag^+(aq) + \frac{1}{2}H_2(g)$	77.1

Apart from being arranged in order of increasing $\Delta G_m^\ominus$, the main characteristic of these reactions is that in all of them, $H^+(aq)$ appears on the left hand side, and $\frac{1}{2}H_2(g)$ on the right. This ensures that when we subtract one equation from another, for example, silver from magnesium, $H^+(aq)$ and $H_2(g)$ are eliminated:

$$\tfrac{1}{2}Mg(s) + Ag^+(aq) = \tfrac{1}{2}Mg^{2+}(aq) + Ag(s); \quad \Delta G_m^\ominus = -304.5 \text{ kJ mol}^{-1} \quad \mathbf{110}$$

The resulting equation, with its negative value of $\Delta G_m^\ominus$, tells us that, thermodynamically, magnesium is less stable than silver with respect to oxidation into its aqueous ion, or, put another way, magnesium is a more powerful reducing agent than silver in aqueous solution. Because the coefficients of $H_2(g)$ and $H^+(aq)$ are the same in all the equations, and because the equations are arranged in order of increasing $\Delta G_m^\ominus$, this conclusion can be reached immediately merely by noting that magnesium lies above silver.

Instead of comparing the values of $\Delta G_m^\ominus$ for such equations, it is more usual to compare values of $E^\ominus$, the standard electrode potential or standard redox potential. The relationship between the two methods of comparison is best established by showing how the relevant $E^\ominus$ values can be obtained from the values of $\Delta G_m^\ominus$.

First we reverse the equations, thus reversing the sign of $\Delta G_m^\ominus$. Then we introduce a convenient shorthand by writing the symbol e in place of what is now $\frac{1}{2}H_2(g)$ on the left and $H^+(aq)$ on the right. Thus, e means $[\frac{1}{2}H_2(g) - H^+(aq)]$ or 'add $\frac{1}{2}H_2(g)$ to the left-hand side and $H^+(aq)$ to the right'. This substitution gives us:

Reaction	$\Delta G_m^\ominus / \text{kJ mol}^{-1}$
$Na^+(aq) + e = Na(s)$	261.9
$\frac{1}{2}Mg^{2+}(aq) + e = \frac{1}{2}Mg(s)$	227.4
$\frac{1}{3}Al^{3+}(aq) + e = \frac{1}{3}Al(s)$	161.7
$Ag^+(aq) + e = Ag(s)$	-77.1

Notice that the $\Delta G_m^\ominus$ values are now arranged in *descending* order. The coefficient of e in all these equations is one, so that e is eliminated when two equations are subtracted. However, we now multiply the equations to eliminate fractions:

Reaction	$\Delta G_m^\ominus / \text{kJ mol}^{-1}$
$Na^+(aq) + e = Na(s)$	261.9
$Mg^{2+}(aq) + 2e = Mg(s)$	454.8
$Al^{3+}(aq) + 3e = Al(s)$	485.1
$Ag^+(aq) + e = Ag(s)$	-77.1

Notice that the equations are no longer arranged in descending order of $\Delta G_m^\ominus$. Order returns, however, when we calculate $E^\ominus$ by using the relationship

$$\Delta G_m^\ominus = -nFE^\ominus \qquad 77$$

Here n is the coefficient of e, and F is a positive constant called the Faraday, which has the value $96\,485\,\text{C}\,\text{mol}^{-1}$, C being the symbol for the coulomb. According to the SI system, a volt (V) is equivalent to a joule per coulomb $(\text{V} = \text{J}\,\text{C}^{-1})$ so $F = 96\,485\,\text{J}\,\text{V}^{-1}\,\text{mol}^{-1}$ or $96.485\,\text{kJ}\,\text{V}^{-1}\,\text{mol}^{-1}$. To obtain $E^\ominus$ from $\Delta G_m^\ominus$, one divides $-\Delta G_m^\ominus/\text{kJ}\,\text{mol}^{-1}$ by $96.485n$. This gives $E^\ominus/\text{V}$. Thus, for aluminium,

$$E^\ominus/\text{V} = -\frac{485.1}{96.485 \times 3}$$

$$E^\ominus = -1.68\,\text{V}$$

The new table becomes:

Reaction	$E^\ominus/\text{V}$
$\text{Na}^+(\text{aq}) + \text{e} = \text{Na(s)}$	-2.71
$\text{Mg}^{2+}(\text{aq}) + 2\text{e} = \text{Mg(s)}$	-2.36
$\text{Al}^{3+}(\text{aq}) + 3\text{e} = \text{Al(s)}$	-1.68
$\text{Ag}^+(\text{aq}) + \text{e} = \text{Ag(s)}$	0.80

The equations are now arranged in ascending order of $E^\ominus$, thus showing that $E^\ominus$, like the particular values of $\Delta G_m^\ominus$ in the first set of equations in this Section, provides a measure of the thermodynamic stability of metals with respect to oxidation to the aqueous cation. Notice that, as a comparison of the second and third sets of equations in this Section shows, $\Delta G_m^\ominus$ in general depends on the coefficient of e; $E^\ominus$, however, does not, because n appears in equation 77. Thus, $E^\ominus = -1.68\,\text{V}$ for all three of the following equations:

Reaction	$\Delta G_m^\ominus/\text{kJ}\,\text{mol}^{-1}$
$\tfrac{1}{3}\text{Al}^{3+}(\text{aq}) + \text{e} = \text{Al(s)}$	161.7 ($n=1$)
$\text{Al}^{3+}(\text{aq}) + 3\text{e} = \text{Al(s)}$	485.1 ($n=3$)
$2\text{Al}^{3+}(\text{aq}) + 6\text{e} = 2\text{Al(s)}$	970.2 ($n=6$)

This explains why $E^\ominus$ is such a convenient measure of the oxidising or reducing powers of redox systems. Although any one of the three equations would do, it is customary to cite the middle one in tables of $E^\ominus$ values, such as that in the S343 *Data Book*. Notice that in this Table, the oxidised state is always written on the left. Each system in the Table is called a *couple*, and some are more complicated than the simple metal–aqueous ion systems considered so far. However, equation 77 holds for each. Thus, for the couple

$$\text{MnO}_4^-(\text{aq}) + 8\text{H}^+(\text{aq}) + 5\text{e} = \text{Mn}^{2+}(\text{aq}) + 4\text{H}_2\text{O(l)}, \quad E^\ominus = 1.51\,\text{V} \qquad 111$$

It follows that

$$\Delta G_m^\ominus = -5 \times 96.485 \times 1.51\,\text{kJ}\,\text{mol}^{-1}$$

$$= -728\,\text{kJ}\,\text{mol}^{-1}$$

Remembering that e stands for $[\tfrac{1}{2}\text{H}_2(\text{g}) - \text{H}^+(\text{aq})]$, this tells us that for the reaction:

$$\text{MnO}_4^-(\text{aq}) + 3\text{H}^+(\text{aq}) + \tfrac{5}{2}\text{H}_2(\text{g}) = \text{Mn}^{2+}(\text{aq}) + 4\text{H}_2\text{O(l)}, \quad \Delta G_m^\ominus = -728\,\text{kJ}\,\text{mol}^{-1} \qquad 112$$

In aqueous solution, any reduced state in a couple is thermodynamically capable of reducing the oxidised state of a couple with a more positive $E^\ominus$. Conversely, any oxidised state in a couple is thermodynamically capable of oxidising the reduced state of a couple with a more negative value of $E^\ominus$. Broadly speaking, couples with $E^\ominus$ values less (more negative) than about $-0.1\,\text{V}$ in acid solution contain powerful reducing agents. Those with $E^\ominus$ values greater (more positive) than about $1.1\,\text{V}$ in acid solution contain powerful oxidising agents.

APPENDIX 2 POTENTIAL DIAGRAMS

This Section should be read in parallel with Section 11.1. It looks into the characteristics of **potential diagrams** such as the one shown below for cobalt:

$$Co^{3+}(aq) \xrightarrow{1.94} Co^{2+}(aq) \xrightarrow{-0.28} Co(s)$$
$$\underset{0.46}{\longrightarrow}$$

The figures over the arrows are the values of $E^\ominus/V$.

> The main purpose of this Section is to establish the fact that, between any two species in any cycle on a potential diagram, the sum of the values of $nE^\ominus$ in a clockwise direction is equal to the sum of the values of $nE^\ominus$ in an anticlockwise direction, where n is the change in oxidation state (always positive) for a particular step.

SLC 19

The proof relies on one of several possible ways of stating the first law of thermodynamics (introduced in a Second Level Course): for a function of state such as the enthalpy or the Gibbs function, the change between any two states of a system depends only on those states and not on how the change was carried out. We shall use this to prove the $nE^\ominus$ rule for the general case, and then apply it to the cobalt example above.

If we take two points on a thermodynamic cycle, such as Co^{3+} and Co in the potential diagram above, our statement of the first law of thermodynamics tells us that whether we move from one to the other by the clockwise route or by the anticlockwise route, the total value of $\Delta G_m^\ominus$ is the same. The total value of $\Delta G_m^\ominus$ by the clockwise route is the sum of the values of $\Delta G_m^\ominus$ for the individual steps; this is written as $\Sigma \Delta G_m^\ominus$(clockwise). Likewise, the total value of $\Delta G_m^\ominus$ by the anticlockwise route is $\Sigma \Delta G_m^\ominus$(anticlockwise). Thus

$$\Sigma \Delta G_m^\ominus(\text{clockwise}) = \Sigma \Delta G_m^\ominus(\text{anticlockwise}) \qquad 113$$

If the cycle involves redox reactions in aqueous solution, then we can substitute by using equation 77:

$$\Sigma - nFE^\ominus(\text{clockwise}) = \Sigma - nFE^\ominus(\text{anticlockwise}) \qquad 114$$

For the individual steps, the values of n may be different but the value of F, a constant, is the same. Thus, $-F$ can be divided out of the term for each step:

$$-F\Sigma nE^\ominus(\text{clockwise}) = -F\Sigma nE^\ominus(\text{anticlockwise}) \qquad 115$$

Cancelling $-F$

$$\Sigma nE^\ominus(\text{clockwise}) = \Sigma nE^\ominus(\text{anticlockwise}) \qquad 116$$

This proves the rule for the general case. For the cobalt example given at the beginning of this Section, the potential diagram is shorthand for the balanced cycle below:

$$Co^{3+}(aq) \xrightarrow{e} Co^{2+}(aq) \xrightarrow{2e} Co(s)$$
$$\underset{3e}{\longrightarrow}$$

Moving from $Co^{3+}(aq)$ to $Co(s)$ by the clockwise route, the first and second steps are:

$$Co^{3+}(aq) + e = Co^{2+}(aq) \qquad 117$$

$$Co^{2+}(aq) + 2e = Co(s) \qquad 118$$

Using equation 77,

$$\Sigma \Delta G_m^\ominus(\text{clockwise}) = -FE^\ominus(\text{Co}^{3+}|\text{Co}^{2+}) - 2FE^\ominus(\text{Co}^{2+}|\text{Co}) \qquad 119$$

By the anticlockwise route there is only one step:

$$\text{Co}^{3+}(\text{aq}) + 3e = \text{Co}(s) \qquad 120$$

$$\Sigma \Delta G_m^\ominus(\text{anticlockwise}) = -3FE^\ominus(\text{Co}^{3+}|\text{Co}) \qquad 121$$

From equation 113,

$$-FE^\ominus(\text{Co}^{3+}|\text{Co}^{2+}) - 2FE^\ominus(\text{Co}^{2+}|\text{Co}) = -3FE^\ominus(\text{Co}^{3+}|\text{Co}) \qquad 122$$

Whence, dividing by $-F$

$$E^\ominus(\text{Co}^{3+}|\text{Co}^{2+}) + 2E^\ominus(\text{Co}^{2+}|\text{Co}) = 3E^\ominus(\text{Co}^{3+}|\text{Co}) \qquad 123$$

It can be seen that the values in the potential diagram at the beginning of this Section bear this out.

OBJECTIVES FOR BLOCK 1

Now that you have completed Block 1, you should be able to do the following things:

1 Recognise valid definitions of, and use in a correct context, the terms, concepts and principles in Table A.

Table A List of scientific terms, concepts and principles used in Block 1

Term	Page No.
Bessemer process	21
bidentate ligand	34
charge-transfer transition	16
chelating agent	34
chloride process for TiO_2 manufacture	9
coordination compound	34
couple	49
d–d transition	15
facial isomer	75
hiding power	10
inner sphere of a complex	32
Kroll process for Ti extraction	12
Leclanché cell	25
meridional isomer	75
outer sphere of a complex	32
polydentate ligand	35
potential diagram	68
standard redox potential	49, 66
sulphate process for TiO_2 manufacture	8
superalloys	29
transition elements	5
unidentate ligand	34
vertical integration	21

2   Derive the electronic configuration of the free ions of transition elements with a charge of +2 or more, and the d-electron population of transition elements in compounds where the transition element has an oxidation state of +2 or more. (SAQs 1, 2, 28 and 34)

3   Recall competing advantages and disadvantages of two industrial processes for the manufacture of TiO_2. (SAQs 3 and 4)

4   Account for the differing costs of aluminium and titanium metals in terms of the chemistry of the extraction processes. (SAQs 5 and 6)

5   Recognise relationships between the structure of a compound, its stoichiometry, its physical properties and the coordination numbers of its component atoms. (SAQs 7, 8 and 11)

6   Recognise the relationship between the colours of chemical species, and the positions of the bands in their visible absorption spectra. (SAQs 9 and 20)

7   Write down probable decomposition reactions for unknown halides and simple aqueous ions of first-row transition elements. (SAQs 10, 15, 16, 23, 40–45 and 47)

8   Recognise likely similarities and likely differences between transition elements and typical elements whose atoms have the same number of outer electrons. (SAQs 12 and 33)

9   Describe and explain the contribution made by metallic manganese to the establishment of the steel industry. (SAQs 13 and 14)

10   Balance redox equations for reactions in both acid and alkaline solution. (SAQs 17, 18 and 51)

11   Recall reactions by which the common oxidation states of titanium, manganese or cobalt in aqueous solution can be interconverted. (SAQs 19, 21 and 22)

12   Given the composition of a complex of chromium(III), cobalt(III) or platinum(IV), and information on the conductivity and chemical behaviour of its aqueous solution, identify the inner and outer spheres of the complex. (SAQs 24–27 and 32)

13   Given the composition of a transition-metal complex, and the coordination exercised by the central element, draw the possible stereoisomers, and identify those that are optically active. (SAQs 29–32)

14   Combine descriptive chemistry in the Block with generalisations 1–17 in Section 14 to predict or rationalise appropriate descriptive chemistry of the transition metals which is not explicitly given in the main text. (SAQs 12, 16, 23, 33–37, 40, 41, 43, 44, 45, 47, 49, 51 and 53–56)

15   Calculate $E^\ominus$ values from $\Delta G_m^\ominus$ values, and use them to determine if a redox reaction is thermodynamically favourable. (SAQs 38–42, 46, 49, 50 and 52)

16   Given a potential diagram containing a cycle with one missing $E^\ominus$ value, calculate the missing value. (SAQs 48 and 49)

17   Describe in terms of standard redox potentials the conditions that determine whether an aqueous ion in acid solution is:

(a)   thermodynamically unstable with respect to reduction by the solvent system;

(b)   thermodynamically unstable with respect to oxidation by the solvent system;

(c)   thermodynamically unstable with respect to disproportionation.

(SAQs 40, 41, 46, 50 and 52)

SAQ ANSWERS AND COMMENTS

SAQ 1
(Objective 2)
Cr^{3+} is [Ar]$3d^3$ and Co^{4+} is [Ar]$3d^5$. Remove the s electrons first and then the d electrons.

SAQ 2
(Objective 2)
The oxidation state is given first, and is followed by the d-electron configuration, which is computed for the monatomic ion with a charge equal to the oxidation state. (a) $+2, 3d^5$; (b) $+6, 3d^1$; (c) $+3, 3d^4$; (d) $+7$, no 3d electrons; (e) $+3, 3d^4$; (f) $+2, 3d^5$.

SAQ 3
(Objective 3)
A, rutile, $TiO_2(s)$; B, $TiCl_4(g)$; C, $O_2(g)$; D, $Cl_2(g)$; E, $TiO_2(s)$. Reactions 1 and 2 are equations 6 and 7 of the main text.

SAQ 4
(Objective 3)
The stoichiometry of equation 8 shows that the production of 1 mole of $TiCl_4$ requires 3 moles of Cl_2, compared with 2 moles of Cl_2 when rutile is used as the starting material. The extra chlorine is used to produce an unwanted by-product, $FeCl_2$. Costs are increased by unnecessary chlorine consumption and the need for $FeCl_2$ disposal. The process could become much more attractive as the price of rutile rises, but this would be especially so if a cheap way of recovering chlorine from $FeCl_2$ could be developed, or if a market could be found for the iron(II) chloride.

SAQ 5
(Objective 4)
SLC 20
The Hall–Heroult process for the extraction of aluminium was discussed in a Second Level Course. Al_2O_3 is dissolved in molten cryolite, Na_3AlF_6, and electrolysed with carbon anodes and a steel cathode at about 950 °C. Molten aluminium sinks to the bottom of the electrolytic cell and is periodically tapped off: the process is continuous. Before the development of the Hall process, aluminium was extracted by methods similar to those now used for titanium: hydrolytically unstable $AlCl_3$ was reduced in batches by an expensive alkali metal such as sodium or potassium.

SAQ 6
(Objective 4)
The very high melting temperature of titanium means that at economical working temperatures, titanium is obtained in a dispersed solid form (sponge) rather than as a liquid, which can then be cast into ingots. Thus, a further *consolidation* step—electric arc melting under vacuum—is necessary. The production of metal in liquid form is also a great help in running a continuous process because it can be tapped off at convenient intervals. Notice that in the electrolytic process of Figure 8, the current flow must be interrupted about every four hours so that the cathode can be lifted and the titanium sponge scraped off.

SAQ 7
(Objective 5)
The boiling temperatures of $TiCl_4$, $TiBr_4$ and TiI_4 increase with molar mass as one might expect in a series of compounds comprising discrete molecules. But at 25 °C, where $TiCl_4$ is a liquid, TiF_4 is a solid, which can only be vaporised at a temperature 150 °C above the boiling temperature of $TiCl_4$. The structure of TiF_4 is not known, but because of these facts, it is assumed to be polymeric, and, possibly, to contain titanium coordinated to six fluorines.

SAQ 8
(Objective 5)
Three; in $TiOSO_4 \cdot H_2O$, there is one sulphate group per titanium, so the coordination number of titanium with respect to sulphate must equal the coordination number of sulphate with respect to titanium. But in Figure 11, each titanium is linked to *three* oxygens of three different sulphate groups. Thus, each sulphate must be linked to three different titaniums in different chains: any one sulphate group links three different chains together.

SAQ 9
(Objective 6)
CS
At lower energies: TiF_3 reflects or transmits light from the blue end of the visible spectrum (see Plate 1.1 of *Colour Sheet 1*), so it absorbs at the red or low-energy end. In fact, the single absorption band has a peak at about 16 000 cm^{-1} compared with 20 000 cm^{-1} for Ti^{3+}(aq). In both cases, the band is due to the d–d transition of Figure 13. Thus, when Ti^{3+} is placed in an environment of fluoride ions, the transition occurs at a lower energy than when it is placed in an environment of water molecules. This kind of difference assumes great importance in Block 2.

SAQ 10
(Objective 7)
The reddish-purple solution presumably contains Ti^{3+}(aq), so the equation is

$$TiCl_2(s) + H^+(aq) = Ti^{3+}(aq) + 2Cl^-(aq) + \tfrac{1}{2}H_2(g)$$

Normally one would expect the dihalide of a metallic element to dissolve as follows:

$$MCl_2(s) = M^{2+}(aq) + 2Cl^-(aq)$$

However, in this case, $Ti^{2+}(aq)$ must be so unstable with respect to oxidation that it then reduces water or $H^+(aq)$:

$$Ti^{2+}(aq) + H^+(aq) = Ti^{3+}(aq) + \tfrac{1}{2}H_2(g)$$

This reaction explains why $Ti^{2+}(aq)$ has not yet been obtained.

SAQ 11 (*Objective 5*) See Figure 41. All titanium atoms in the deck lie on the set of lines with the direction and spacing specified by arrows. Along each line there is a repeat pattern of two titaniums due to stay followed by one due to be removed. Thus, the proportion eliminated from $TiCl_2$ is one-third, and two-thirds or 0.67 remain. This gives a stoichiometry $Ti_{0.67}Cl_2$, which is equivalent to $TiCl_3$.

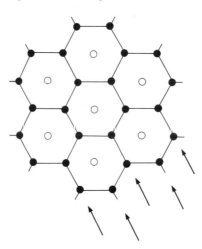

Figure 41 The pattern of titanium removal from the titanium layers in $TiCl_2$ which generates $TiCl_3$.

SAQ 12 (*Objectives 8 and 14*) Points (a)–(c) encourage the classification, because they are equally true for titanium. Points 4 and 5 are not favourable to the classification because in these properties there are characteristic differences between a typical element like tin and a transition element like titanium. Thus, titanium has a very high melting temperature (Section 3.4) and forms compounds in a set of oxidation states ($+2$, $+3$, $+4$) which differ by one (Section 3.6.3).

SAQ 13 (*Objective 9*) The difference is that the oxidation of manganese is generally faster and thermodynamically more favourable than that of iron. Thus, manganese rapidly takes up dissolved oxygen in molten iron. Likewise, in the presence of sulphur, manganese(II) sulphide rather than iron(II) sulphide is formed. Notice, however, that the process also relies on liquid manganese being *soluble* in the liquid iron: this gives it access to the impurities.

SAQ 14 (*Objective 9*) $\Delta G_m^\ominus = -118$ kJ mol^{-1}. Because the values of $\Delta G_f^\ominus(Mn, s)$ and $\Delta G_f^\ominus(Fe, s)$ are zero by convention,

$$\Delta G_m^\ominus = \Delta G_f^\ominus(MnS, s) - \Delta G_f^\ominus(FeS, s)$$

The negative sign shows that the reaction is thermodynamically favourable. Thermodynamics suggests that when iron and manganese compete for sulphur, manganese wins. This is consistent with the role of manganese as a sulphur scavenger in steelmaking.

SAQ 15 (*Objective 7*) The only known manganese chloride is $MnCl_2$, so it is reasonable to suppose that $MnCl_3$ is unstable with respect to it:

$$MnCl_3(s) = MnCl_2(s) + \tfrac{1}{2}Cl_2(g)$$

SAQ 16 (*Objectives 7 and 14*) For titanium, all four trihalides are known, and all dihalides except TiF_2. For manganese, MnF_3 is the only trihalide that exists at room temperature, but all the dihalides can be prepared. The stability of trihalides with respect to dihalides is presumably controlled by reactions of the type discussed in SAQ 15:

$$MX_3(s) = MX_2(s) + \tfrac{1}{2}X_2$$

The range of known trihalides suggests that the trihalides of manganese are much less stable with respect to this reaction than are those of titanium. As MnF_3 is the only manganese trihalide stable at room temperature, fluorine seems the best halogen for stabilising the higher of the two oxidation states.

SAQ 17 (*Objective 10*) The required equation is

$$5NaBiO_3(s) + 2Mn^{2+}(aq) + 14H^+(aq) = 2MnO_4^-(aq) + 5Bi^{3+}(aq) + 5Na^+(aq) + 7H_2O(l)$$

SLC 8 In a Second Level Course, you were taught both the *oxidation state* method, and the *ion–electron* method of balancing redox equations. Here, we shall use the oxidation state method.

Bismuth is reduced from oxidation state $+5$ in $NaBiO_3$, to $+3$ in Bi^{3+}(aq), a change of -2; manganese is oxidised from $+2$ in Mn^{2+}(aq) to $+7$ in MnO_4^-(aq), a change of $+5$. In a balanced equation the total change in oxidation state must be zero, so 5 moles of $NaBiO_3$(s) must react with 2 moles of Mn^{2+}(aq) because this gives a total change of $(-10 + 10)$. The balanced equation will therefore include the combination:

$$5NaBiO_3(s) + 2Mn^{2+}(aq) \longrightarrow 5Na^+(aq) + 5Bi^{3+}(aq) + 2MnO_4^-(aq)$$

One now adds $7H_2O$(l) to the right to balance oxygen. This leaves a deficiency of hydrogen on the left, which, as the reaction occurs in acid, can be redressed by adding $14H^+$(aq) to that side to give the final answer. The equation has now been balanced with respect to chemical elements, and if the job has been correctly done there is a final check: it should be balanced with respect to charge. And it is: the charges sum to $+18$ on each side.

SAQ 18
(Objective 10)

(a) $2MnO_4^{2-}(aq) + SO_3^{2-}(aq) + 2OH^-(aq) = 2MnO_4^{3-}(aq) + SO_4^{2-}(aq) + H_2O(l)$

Manganese is reduced from oxidation state $+6$ to $+5$; sulphur is oxidised from $+4$ to $+6$, so the answer will contain the combination

$$2MnO_4^{2-}(aq) + SO_3^{2-}(aq) \longrightarrow 2MnO_4^{3-}(aq) + SO_4^{2-}(aq)$$

With equations in alkali where one must use OH^- ions in the balancing act, it is best to balance charges with OH^- at the next stage. With a sum of -6 on the left, and -8 on the right, $2OH^-$ must be added to the left. By adding H_2O to the right, the hydrogen imbalance is corrected. A final check on the answer is provided by an oxygen inventory: there are 13 oxygens on each side.

(b) $3MnO_4^{3-}(aq) + 8H^+(aq) = MnO_4^-(aq) + 2MnO_2(s) + 4H_2O(l)$

Manganese is simultaneously oxidised from oxidation state $+5$ to $+7$, and reduced from $+5$ to $+4$. Therefore, $3MnO_4^{3-}$ must disproportionate to $2MnO_2$ and one MnO_4^-. Hydrogen and oxygen are then balanced as in SAQ 17.

SAQ 19
(Objective 11)

If equation 30 is regarded as an equilibrium system, Le Chatelier's principle predicts that Mn^{3+}(aq) will be stabilised at high concentrations of H^+(aq). This is correct: the stability of aqueous acidic solutions of manganese(III) increases as the concentration of the acid increases.

SAQ 20
(Objective 6)

Plate 1.1 of *Colour Sheet 1* shows that the visible spectrum roughly spans 14 000–25 000 cm^{-1}. The colour of MnO_4^-(aq) is mainly due to the broad absorption band with a peak at about 19 000 cm^{-1}. This removes the central part of the visible spectrum, leaving a mix of red and blue, which accounts for the purple colour of permanganate. MnO_4^{2-}(aq) absorbs strongly through nearly all the visible region, except for a window centred near 19 000 cm^{-1} in the green; hence solutions of manganate(VI) are green. MnO_4^{3-}(aq) absorbs visible light mainly between 13 000 and 17 000 cm^{-1} due to the high-energy end of a broad absorption band, with a peak near 15 000 cm^{-1}. The blue-green end of the spectrum from 17 000 cm^{-1} upwards is transmitted; this is the colour of MnO_4^{3-}(aq).

SAQ 21
(Objective 11)

A $BaMnO_4$(s); B MnO_2(s); C MnO_4^-(aq); D Mn^{2+}(aq); E O_2(g); F Mn^{3+}(aq). The green colour of A suggests that it contains the ion MnO_4^{2-}, and the barium salt of such an ion is $BaMnO_4$. Note that it must be formed by reduction of MnO_4^- in strong alkali:

$$2MnO_4^-(aq) + 2OH^-(aq) = 2MnO_4^{2-}(aq) + \tfrac{1}{2}O_2(g) + H_2O(l)$$

Then follows disproportionation of A (equation 36), dissolution and reduction of B (equation 42), and anodic oxidation of pale-pink Mn^{2+}(aq) to red Mn^{3+}(aq).

SAQ 22
(Objective 11)

They are somewhat similar. In Section 4.4, you saw that oxygen does not oxidise Mn^{2+}(aq) to Mn^{3+}(aq) in acid. In alkali, however, oxygen oxidises $Mn(OH)_2$ to manganese(III) in the compound $MnO(OH)$(s). The chief difference appears in acid solution: unlike Co^{3+}(aq), Mn^{3+}(aq) seems to oxidise water either very slowly or not at all.

SAQ 23
(Objectives 7 and 14)

The range of known dihalides and trihalides is the same for manganese and cobalt: all dihalides plus a trifluoride. MnF_3 sublimes at a temperature above that at which CoF_3 decomposes to CoF_2. This suggests that the stability of the trifluoride is lower for cobalt. SAQ 16 suggested that the stability of the trihalides with respect to the dihalides decreased from titanium to manganese. This new information hints that the trend continues further across the transition series to cobalt.

SAQ 24
(Objective 12)
To obtain a coordination number of six, two of the three chlorines must occupy inner-sphere positions. This gives $[Co(NH_3)_4Cl_2]Cl$ and the dissociation reaction

$$[Co(NH_3)_4Cl_2]Cl(s) = [Co(NH_3)_4Cl_2]^+(aq) + Cl^-(aq)$$

This agrees with the observed fact that one-third of the chloride is precipitable.

SAQ 25
(Objective 12)
$[Cr(NH_3)_4(H_2O)Cl]Cl_2$; the dissolution is

$$[Cr(NH_3)_4(H_2O)Cl]Cl_2(s) = [Cr(NH_3)_4(H_2O)Cl]^{2+}(aq) + 2Cl^-(aq)$$

A coordination number of six is achieved by assuming that H_2O, like NH_3, can occupy an inner-sphere position. The dissolution then gives three ions, two of which are chlorides, and this is consistent with the molar conductivity (Table 10), and with the precipitation of two-thirds of the chlorine.

SAQ 26
(Objective 12)
One compound is $[Co(NH_3)_5(SO_4)]Cl$, whose solution will give a precipitate of silver chloride with silver nitrate. The other is $[Co(NH_3)_5Cl]SO_4$, whose solution will give a precipitate of barium sulphate with barium chloride.

SAQ 27
(Objective 12)
A platinum coordination number of four is common to all the formulae $[Pt(NH_3)_4]Cl_2$, $[Pt(NH_3)_3Cl]Cl$, $[Pt(NH_3)_2Cl_2]$, $K[Pt(NH_3)Cl_3]$ and $K_2[PtCl_4]$. In aqueous solution, these compounds will yield 3, 2, 0, 2 and 3 ions, respectively. For example:

$$[Pt(NH_3)_4]Cl_2(s) = [Pt(NH_3)_4]^{2+}(aq) + 2Cl^-(aq)$$

Reference to Table 10 shows that these numbers of ions are consistent with the molar conductivities in Table 11.

SAQ 28
(Objective 2)
The oxidation states are chromium(III) and nickel(II); they are associated with the d-electron configurations $3d^3$ and $3d^8$, respectively. For $[Cr(NH_3)Cl(en)_2]Cl_2$, the closed-shell configurations of the ligands are en, NH_3 and Cl^-. Removal of 2en, NH_3 and $3Cl^-$ leaves Cr^{3+}, which has the configuration $[Ar]3d^3$. With $[Ni(edta)]^{2-}$, the closed-shell configuration of the ligand is $edta^{4-}$. Its removal leaves Ni^{2+}, with the configuration $[Ar]3d^8$.

SAQ 29
(Objective 13)
Square planar: in tetrahedral coordination, only one compound $[Pt(NH_3)_2Cl_2]$ would exist, just as there is only one stereoisomer of CH_2Cl_2. In square-planar coordination there are *cis* (structure **12**) and *trans* (structure **13**) forms, both of which have planes of symmetry and are therefore optically inactive:

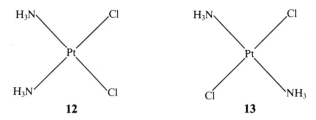

SAQ 30
(Objective 13)
There are two optical isomers (structures **14** and **15**), which are non-superimposable mirror images of each other:

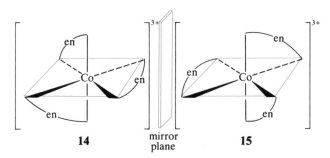

Neither complex has a plane of symmetry.

SAQ 31
(Objective 13)
See Figures 42 and 43. For $[CoBr_2Cl_2(en)]^-$ there are four stereoisomers, two of which (C and D) are optically active. For $[Co(NH_3)_3(H_2O)BrCl]^+$, there are five stereoisomers, two of which (D and E) are optically active.

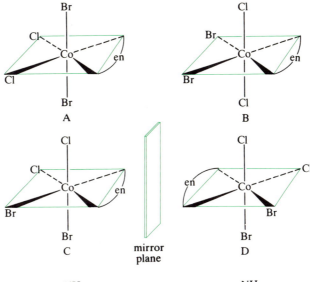

Figure 42 Stereoisomers of [CoBr$_2$Cl$_2$(en)]$^-$.

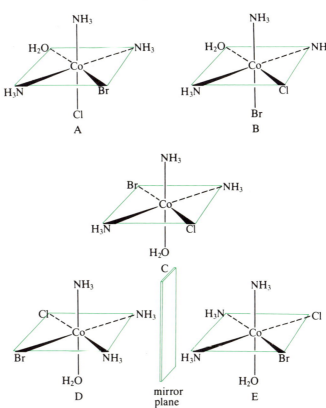

Figure 43 Stereoisomers of [Co(NH$_3$)$_3$(H$_2$O)BrCl]$^+$.

SAQ 32
(Objectives 12 and 13)

The compound is [CoCl$_3$(dien)], and there are just two stereoisomers, neither of which is optically active. As dien is tridentate (Table 13), two dien ligands would occupy all six positions and push the chloride into the outer sphere, where it would be precipitated by silver nitrate solution. There must therefore be one dien and three chlorides in the inner sphere. As the substance is a cobalt(III) complex, the result is the neutral molecular complex [CoCl$_3$(dien)]. As the central —NH— group of dien must be adjacent to both terminal —NH$_2$ groups, two arrangements are possible. In the **facial isomer, 16** (denoted *fac*-[CoCl$_3$(dien)]) the three nitrogens occupy the corners of one face of the octahedron. In the **meridional isomer, 17** (denoted *mer*-[CoCl$_3$(dien)]) they occupy three positions on a meridian of the octahedron. Both stereoisomers have planes of symmetry so neither is optically active.

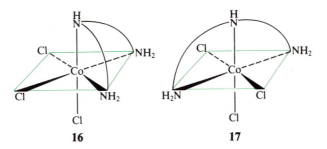

SAQ 33
(Objectives 8 and 14)

A is zirconium and B is tin. The elements with four valence electrons in Period 5 are zirconium, [Kr]$4d^25s^2$, and tin, [Kr]$4d^{10}5s^25p^2$. As expected from the conductivity evidence, both are metals. A has much the higher melting temperature so it must be zirconium, a transition metal.

SAQ 34
(Objectives 2 and 14)

Zn^{2+}(aq) is associated with the configuration [Ar]$3d^{10}$. In Section 3.6.2 we said that the d–d transitions within incomplete d shells were an important source of colour in transition-metal chemistry. In Zn^{2+}(aq) the 3d shell is full, so these transitions cannot occur and the ion is colourless.

SAQ 35
(Objective 14)

According to generalisations made in Section 8.2.1, NiF_2 has a rutile structure and $FeBr_2$ has a CdX_2 layer structure. In both these structures the metal is octahedrally coordinated by halogen. In NiF_2 (rutile, Figure 3) the fluorine is coordinated to three nickels at the corners of a triangle. In $FeBr_2$ (layer structure, Figure 14) the bromine has three irons on one side of it, and on the other side there are more distant bromines in an adjacent three-deck layer.

SAQ 36
(Objective 14)

This is a difficult question to answer. The general decrease in $r(M^{2+})$ across the series suggests that the radii of Sc^{2+} and Ti^{2+} should be between those of Ca^{2+} and V^{2+}. VF_2 and other transition-metal difluorides have a rutile structure, but CaF_2 has the fluorite structure. Thus, ScF_2 and TiF_2 might have either the rutile or fluorite structure, but as we expect $r(Sc^{2+})$ to be greater than $r(Ti^{2+})$, the probability of ScF_2 having the fluorite structure is greater than for TiF_2.

SAQ 37
(Objective 14)

Suppose that the stabilities of the M^{2+}(aq) and M^{3+}(aq) ions with respect to each other determine the question of the existence or non-existence of M^{2+}(aq) and M^{3+}(aq) for each element. In that case, oxidation of M^{2+}(aq) is so easy at the beginning of the series that M^{2+}(aq) is unknown; on the other hand it is very hard at the end, where M^{3+}(aq) is unknown. It looks as though the job of oxidising M^{2+}(aq) to M^{3+}(aq) becomes harder as we move from scandium to zinc.

SAQ 38
(Objective 15)

$E^\ominus(Mn^{3+}|Mn^{2+}) = 1.60$ V. From Table 19, $\Delta G_m^\ominus(74) = 154$ kJ mol^{-1}. We can then use equation 80 to obtain $E^\ominus$/V, simply dividing 154 by 96.485, the latter constant having been given earlier in the text. Let us, however, use the more general argument. The quoted value of $\Delta G_m^\ominus(74)$ tells us that for

$$Mn^{2+}(aq) + H^+(aq) = Mn^{3+}(aq) + \tfrac{1}{2}H_2(g), \quad \Delta G_m^\ominus = 154 \text{ kJ mol}^{-1}$$

So, reversing the equation to get the oxidised state on the left, we have

$$Mn^{3+}(aq) + \tfrac{1}{2}H_2(g) = Mn^{2+}(aq) + H^+(aq)$$

or $\quad Mn^{3+}(aq) + e = Mn^{2+}(aq)$

for which $\Delta G_m^\ominus = -154$ kJ mol^{-1}.

Using the general equation 77,

$$-154 \text{ kJ mol}^{-1} = -nFE^\ominus$$

But $n = 1$ and, as noted in Appendix 1, the Faraday constant, F, is 96.485 kJ V^{-1} mol^{-1}. This gives $E^\ominus = 1.60$ V.

SAQ 39
(Objective 15)

(a) Favourable; $E^\ominus(Mn^{3+}|Mn^{2+}) > E^\ominus(Cr^{3+}|Cr^{2+})$;

(b) unfavourable; $E^\ominus(Fe^{3+}|Fe^{2+}) < E^\ominus(Mn^{3+}|Mn^{2+})$;

(c) favourable; $E^\ominus(Co^{3+}|Co^{2+}) > E^\ominus(Mn^{3+}|Mn^{2+})$;

(d) favourable; this is oxidation of Fe^{2+}(aq) by Co^{3+}(aq), and $E^\ominus(Co^{3+}|Co^{2+}) > E^\ominus(Fe^{3+}|Fe^{2+})$;

(e) favourable; $E^\ominus(S_2O_8^{2-}|2SO_4^{2-}) > E^\ominus(Mn^{3+}|Mn^{2+})$;

(f) favourable; $E^\ominus(S_2O_8^{2-}|2SO_4^{2-}) > E^\ominus(V^{3+}|V^{2+})$.

SAQ 40
(Objectives 7, 14, 15 and 17)

Ti^{2+}(aq) is unknown (Table 15) and since the stability of M^{3+}(aq) is greater than M^{2+}(aq) at the start of the series, its non-existence is presumably due to its reduction of H^+(aq):

$$Ti^{2+}(aq) + H^+(aq) = Ti^{3+}(aq) + \tfrac{1}{2}H_2(g)$$

According to Table 19 this reaction is thermodynamically favourable because $E^{\ominus}(\text{Ti}^{3+}|\text{Ti}^{2+})$ is negative. If we imagine solid TiCl_2 first behaving like other dihalides, so that it forms $M^{2+}(\text{aq})$ and $2\text{Cl}^-(\text{aq})$ in water, the total reaction is

$$\text{TiCl}_2(\text{s}) + \text{H}^+(\text{aq}) = \text{Ti}^{3+}(\text{aq}) + 2\text{Cl}^-(\text{aq}) + \tfrac{1}{2}\text{H}_2(\text{g})$$

and the black powder should dissolve, evolving H_2 and giving a purple solution of $\text{Ti}^{3+}(\text{aq})$. This does indeed happen. The same conclusion was reached, without any thermodynamic argument, in SAQ 10.

SAQ 41
(Objectives 7, 14, 15 and 17)

$\text{Ni}^{3+}(\text{aq})$ is not known (Table 18) and the trend in the relative stabilities of $M^{2+}(\text{aq})$ and $M^{3+}(\text{aq})$ across the transition series suggests that it is even less stable than $\text{Co}^{3+}(\text{aq})$ with respect to reduction. Presumably, therefore, its non-existence is due to oxidation of water, which already occurs at $\text{Co}^{3+}(\text{aq})$:

$$2\text{Ni}^{3+}(\text{aq}) + \text{H}_2\text{O}(\text{l}) = 2\text{Ni}^{2+}(\text{aq}) + 2\text{H}^+(\text{aq}) + \tfrac{1}{2}\text{O}_2(\text{g})$$

According to Table 19, this reaction is thermodynamically favourable because $E^{\ominus}(\text{Ni}^{3+}|\text{Ni}^{2+}) > 1.23$ V. Thus, when the hydroxide is dissolved in acid, oxygen should be evolved and the green colour of $\text{Ni}^{2+}(\text{aq})$ should appear:

$$2\text{NiO(OH)} + 4\text{H}^+(\text{aq}) = 2\text{Ni}^{2+}(\text{aq}) + 3\text{H}_2\text{O}(\text{l}) + \tfrac{1}{2}\text{O}_2(\text{g})$$

This is what happens in practice.

SAQ 42
(Objectives 7 and 15)

$E^{\ominus}(\text{Cr}^{3+}|\text{Cr}^{2+})$ and $E^{\ominus}(\text{Fe}^{3+}|\text{Fe}^{2+})$ are less than 1.23 V, so $\text{Cr}^{2+}(\text{aq})$ and $\text{Fe}^{2+}(\text{aq})$ are thermodynamically unstable to oxidation by oxygen in acid solution; $E^{\ominus}(\text{Mn}^{3+}|\text{Mn}^{2+})$ and $E^{\ominus}(\text{Co}^{3+}|\text{Co}^{2+})$ are greater than 1.23 V, so $\text{Mn}^{2+}(\text{aq})$ and $\text{Co}^{2+}(\text{aq})$ are thermodynamically stable to this reaction. The oxidation of $\text{Cr}^{2+}(\text{aq})$ by oxygen is rapid at room temperature (Section 9.1), but that of $\text{Fe}^{2+}(\text{aq})$ is very slow, so in laboratory experiments like those shown in Figure 34, solutions of $\text{Fe}^{2+}(\text{aq})$ can be handled in air without serious deterioration. Oxidation of $\text{Co}^{2+}(\text{aq})$ and $\text{Mn}^{2+}(\text{aq})$ by oxygen in acid solution never happens, of course, because it is thermodynamically unfavourable. Thus, predictions from thermodynamic data alone are only seriously misleading for $\text{Fe}^{2+}(\text{aq})$.

SAQ 43
(Objectives 7 and 14)

MnCl_3 is stable below $-40\,°\text{C}$, but MnBr_3 and MnI_3 are unknown, and we presume this is because they are even less stable with respect to the dihalides. Likewise, FeCl_3 is known but FeI_3 is not.

SAQ 44
(Objectives 7 and 14)

Applying the stability trend observed in the trihalides, the oxides M_2O_3 should become less stable with respect to the reaction

$$\tfrac{1}{2}M_2O_3(\text{s}) = MO(\text{s}) + \tfrac{1}{4}O_2(\text{g})$$

as one moves across the series, so the M^{III} oxides of the first six elements (Sc_2O_3, Ti_2O_3, V_2O_3, Cr_2O_3, Mn_2O_3, Fe_2O_3) are a good guess. The next one, Co_2O_3, should be the least stable, and would be expected to lose oxygen at lower temperatures. This is correct. Like fluorine, oxygen stabilises seven elements in the tripositive oxidation state and is very good generally at bringing out high oxidation states. Notice the crucial assumption used throughout: the stability trend observed in halides and aqueous ions is transferable to oxides in the same oxidation state.

SAQ 45
(Objectives 7 and 14)

The existence of the insoluble compound CuI described in Section 11.1 suggests the decomposition reaction

$$\text{CuI}_2(\text{s}) = \text{CuI}(\text{s}) + \tfrac{1}{2}\text{I}_2(\text{s})$$

Notice that a solution of CuI_2, made by mixing solutions of $\text{Cu}^{2+}(\text{aq})$ and $\text{I}^-(\text{aq})$, decomposes in this way (equation 97).

SAQ 46
(Objectives 15 and 17)

$E^{\ominus}(\text{Cu}^{2+}|\text{Cu})$ and $E^{\ominus}(\text{Cu}^+|\text{Cu})$ are both positive, so the reactions

$$\text{Cu}(\text{s}) + \text{H}^+(\text{aq}) = \text{Cu}^+(\text{aq}) + \tfrac{1}{2}\text{H}_2(\text{g})$$

$$\text{Cu}(\text{s}) + 2\text{H}^+(\text{aq}) = \text{Cu}^{2+}(\text{aq}) + \text{H}_2(\text{g})$$

are both thermodynamically unfavourable.

SAQ 47
(Objectives 7 and 14)

If prepared, these compounds would probably decompose quickly by disproportionation to the metal and the known oxidation state +2:

$$Ni_2O(s) = Ni(s) + NiO(s)$$

$$2VI(s) = V(s) + VI_2(s)$$

SAQ 48
(Objective 16)

Using the lower left-hand cycle, the values of n for the MnO_4^-/MnO_4^{2-} and MnO_4^{2-}/MnO_2 steps are 1 and 2, respectively, because the oxidation state of manganese decreases by 1 in the first step and by 2 in the second. In the MnO_4^-/MnO_2 step, $n = 3$, because the oxidation state decreases from 7 to 4. Thus

$$(1 \times 0.56)\,V + 2E^\ominus(MnO_4^{2-}\,|\,MnO_2) = 3 \times 1.70\,V$$

$$2E^\ominus(MnO_4^{2-}\,|\,MnO_2) = (5.10 - 0.56)\,V = 4.54\,V$$

$$E^\ominus(MnO_4^{2-}\,|\,MnO_2) = 2.27\,V$$

From the lower right-hand cycle:

$$E^\ominus(MnO_2\,|\,Mn^{3+}) + (1 \times 1.60)\,V = (2 \times 1.23)\,V$$

$$E^\ominus(MnO_2\,|\,Mn^{3+}) = (2.46 - 1.60)\,V$$

$$= 0.86\,V$$

SAQ 49
(Objectives 14, 15 and 16)

Yes: $E^\ominus(S_2O_8^{2-}\,|\,2SO_4^{2-})$, is greater than both $E^\ominus(Mn^{3+}\,|\,Mn^{2+})$ and $E^\ominus(MnO_4^-\,|\,Mn^{3+})$. $E^\ominus(Mn^{3+}\,|\,Mn^{2+}) = 1.60$ V from the potential diagram. To find $E^\ominus(MnO_4^-\,|\,Mn^{3+})$, you use the value of $E^\ominus(MnO_4^-\,|\,MnO_2)$ in the potential diagram (1.70 V) and the value of $E^\ominus(MnO_2\,|\,Mn^{3+})$ calculated in SAQ 48 (0.86 V). Then

$$4E^\ominus(MnO_4^-\,|\,Mn^{3+}) = (3 \times 1.70\,V) + 0.86\,V$$

$$= 5.96\,V$$

$$E^\ominus(MnO_4^-\,|\,Mn^{3+}) = 1.49\,V$$

This is much less than 1.94 V.

SAQ 50
(Objectives 15 and 17)

In both instances, an intermediate oxidation state decomposes to a higher and a lower oxidation state. In the first case, manganese(III) *disproportionates* to manganese(II) and manganese(IV); in the second, manganese(VI) disproportionates to manganese(IV) and manganese(VII). In both cases, the $E^\ominus$ value linking the higher and intermediate oxidation states is less than that linking the intermediate and lower: $E^\ominus(MnO_2\,|\,Mn^{3+}) < E^\ominus(Mn^{3+}\,|\,Mn^{2+})$ and $E^\ominus(MnO_4^-\,|\,MnO_4^{2-}) < E^\ominus(MnO_4^{2-}\,|\,MnO_2)$.

SAQ 51
(Objectives 10 and 14)

Yes. Chlorine oxidises iron(III) hydroxide to FeO_4^{2-} in alkaline solution, and under these conditions the iron(VI) that is formed is stable with respect to reduction. However, when the iron(VI) compound is added to acid, FeO_4^{2-} decomposes to a yellow–brown solution of Fe^{3+}(aq), some of the oxygen from the oxoanion being converted to molecular oxygen. The higher oxidation state is thus more stable in alkaline solution. The equation for the decomposition of FeO_4^{2-} in acid is consistent with this, H^+(aq) appearing on the left-hand side:

$$2FeO_4^{2-}(aq) + 10H^+(aq) = 2Fe^{3+}(aq) + 5H_2O(l) + \tfrac{3}{2}O_2(g)$$

SAQ 52
(Objectives 15 and 17)

(a) Consistent because $E^\ominus(FeO_4^{2-}\,|\,Fe^{3+})$ exceeds 1.23 V.

(b) Consistent because $E^\ominus(Fe^{3+}\,|\,Fe^{2+})$ is less than 1.23 V.

(c) This observation is concerned with the rate of reaction. Thermodynamics, and therefore the potential diagram, can tell us nothing about this.

(d) Consistent because $E^\ominus(Fe^{2+}\,|\,Fe)$ is negative. Compare SAQ 46.

(e) Inconsistent in the sense that, because $E^\ominus(Fe^{3+}\,|\,Fe^{2+})$ is positive and hence greater than $E^\ominus(H^+\,|\,\tfrac{1}{2}H_2)$, the reduction of Fe^{3+} to Fe^{2+} by hydrogen gas is thermodynamically favourable. The reaction fails to occur for kinetic reasons.

(f) Consistent because $E^\ominus(Fe^{2+}\,|\,Fe) < E^\ominus(Fe^{3+}\,|\,Fe^{2+})$. If Fe^{2+}(aq) were unstable with respect to disproportionation, the reverse would be true.

SAQ 53
(Objective 14)

If one applies the procedure recommended in the Second Level Course to CrO_4^{2-}, one adds the two electrons of the ion's negative charge to the six valence electrons of the chromium atom ($3d^5 4s^1$). This gives a total of eight electrons, all of which are used in forming four double bonds to oxygen. There are therefore just four repulsion axes, which take up a regular tetrahedral shape (structure **18**).

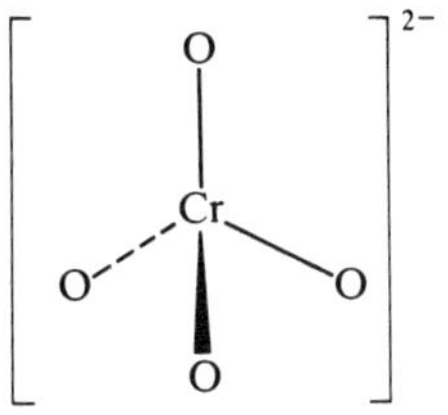

$$\left[\begin{array}{c} O \\ \| \\ O\text{---}Cr\text{---}O \\ \| \\ O \end{array} \right]^{2-} \quad \left[\begin{array}{c} O \\ \| \\ \bullet\bullet\text{---}Fe\text{---}O \\ \diagdown \\ O \end{array} \right]^{2-}$$

18 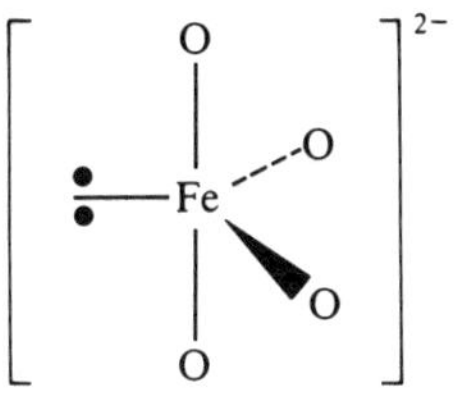 **19**

With FeO_4^{2-}, there are eight outer electrons ($3d^6 4s^2$), so addition of the ion charge gives a pool of ten, and as before, eight are required to form four double bonds. Following orthodox valence-shell electron-pair repulsion theory, this leaves a lone pair, so there are five repulsion axes. The predicted shape is based on a trigonal bipyramid, possibly with the lone pair occupying an equatorial position (structure **19**). In fact, CrO_4^{2-} and FeO_4^{2-} are *both* regular tetrahedral. Indeed, with FeO_4^{2-}, better agreement with the experimental results is achieved by ignoring repulsion from the non-bonding pair of electrons, and assuming that the shape is dictated just by the repulsions between the electrons in the four double bonds.

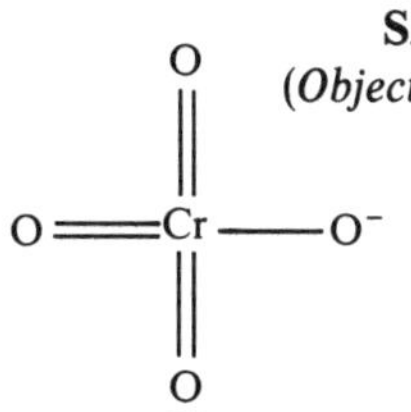

SAQ 54
(Objective 14)

$$O=Cr\text{---}O^-$$
with additional $=O$ groups above and below

20

One resonance structure of the hypothetical anion CrO_4^- is shown in structure **20**. The chromium atom forms one single bond and three double bonds, so it must supply seven electrons. The six valence electrons of the configuration $[Ar]3d^5 4s^1$ are insufficient, so the additional one must come from the inner argon core. It would have to be promoted to the outer 3d, 4s or 4p valence orbitals of chromium, where it could be shared with the surrounding oxygen atoms. This promotion requires so much energy, that if CrO_4^- were to be formed, it would be a very strong oxidising agent, easily reduced to ions like CrO_4^{2-} in which the chromium valence electrons *are* sufficient to form the necessary bonds.

SAQ 55
(Objective 14)

For K_2VO_4 the answer is (a). For K_2NiO_4, Fe_2O_7 and NiF_5, the oxidation state is higher than the maximum given in Figure 40, but does not exceed the number of valence electrons, so the answer is (b). For $ScCl_2$ and TiF_2 the oxidation state is low and the answer is (c).

SAQ 56
(Objective 14)

Certainly, generalisations 1, 2, 3, 4, 11, 12, 13, 15, and 17. Generalisation 15 is applicable because VCl_4 is a regular tetrahedral shape, even though it has one non-bonding valence electron, and 17 because VF_5 (but not VCl_5, VBr_5 and VI_5), and VF_4, VCl_4 and VBr_4 (but not VI_4) are known.

There is a sense in which 5 is in evidence because V^{2+} is readily oxidised in air. The maximum oxidation state mentioned is +5 in VO_2^+ and VF_5, illustrating 6, and failure to produce the oxidation state +1 by reduction is in accord with 7.

ACKNOWLEDGEMENTS

Grateful acknowledgement is made to the following sources for material used in this Block.

Figure 1: Ann Ronan Picture Library; *Figure 7*: R. R. Meyers and J. S. Long, *Pigments: Part I*, Vol. 3 of *Treatise on Coatings*, Marcel Dekker, Inc. N.Y. (1975). Reprinted from Figure 10.5, p. 489 by courtesy of Marcel Dekker, Inc.; *Figure 9*: *Jane's All the World's Aircraft 1974/75*, Jane's Publishing Co. Ltd, (1974); *Figure 18*: E. R. Toon, G. L. Ellis and J. Brodkin, *Foundations of Chemistry*, Holt, Rinehart and Winston, Inc. (1968); *Figure 23*: Professor R. K. Sorem of the Marine Mineral Resources Museum, Washington; *Figure 26*: *Les Prix Nobel*, 1913.

S343 Inorganic Chemistry

Block 1 Introducing the transition elements

Block 2 Theory of metal-ligand interaction

Block 3 Transition-metal chemistry: the stabilities of oxidation states

Block 4 Structure, geometry and synthesis of transition-metal complexes

Block 5 Nuclear magnetic resonance spectroscopy

Block 6 Organometallic chemistry

Block 7 Bioinorganic chemistry

Block 8 Solid-state chemistry

Block 9 Actinide chemistry and the nuclear fuel cycle